B.K. Ridley

SEMICONDUCTOR QUANTUM DOTS

SERIES ON ATOMIC, MOLECULAR AND OPTICAL PHYSICS

Series Advisors: Raymond Y. Chiao, Steven Chu, C. Cohen-Tannoudji, S. Haroche, T. Y. Li, L. Mandel, F. de Martini, G. W. Series, Y. Z. Wang, E. Yablonovitch

Vol. 1: Atoms in Electromagnetic Fields
C. Cohen-Tannoudji

WORLD SCIENTIFIC SERIES ON
ATOMIC, MOLECULAR AND OPTICAL PHYSICS
Vol. 2

SEMICONDUCTOR QUANTUM DOTS

L. Bányai

Institut f. Theoretische Physik
Universität Frankfurt, Robert Mayer Str. 8
6000 Frankfurt, Main, Germany

S. W. Koch

Optical Sciences Center and
Department of Physics, University of Arizona
Tucson, AZ 85721, U.S.A

World Scientific
Singapore • New Jersey • London • Hong Kong

Published by

World Scientific Publishing Co. Pte. Ltd.
P O Box 128, Farrer Road, Singapore 9128
USA office: Suite 1B, 1060 Main Street, River Edge, NJ 07661
UK office: 73 Lynton Mead, Totteridge, London N20 8DH

The authors and publisher are grateful for permission to reproduce the reprinted figures found in this volume.

Series on Atomic, Molecular and Optical Physics – Vol. 2
SEMICONDUCTOR QUANTUM DOTS

ISSN   0218-5032
ISBN   981-02-1390-5

Printed in Singapore.

PREFACE

This book presents an overview of the theoretical background and recent developments in the rapidly growing field of ultrasmall semiconductor microcrystallites, in which carrier confinement in all spatial dimensions plays an important role. The main emphasis of this book is the theoretical analysis of the confinement induced modifications of the optical and electronic properties of quantum dots in comparison to extended (bulk) semiconductors. This book develops the theoretical background material for the analysis of carrier quantum confinement effects, it introduces different quantization regimes, and it gives an overview of the relevant approximation schemes for each regime. A detailed discussion of the carrier states in quantum dots is presented, including variational calculations, a configuration interaction approach, and Quantum Monte Carlo calculations. Surface polarization effects are analyzed which lead to self-trapping of carriers near the surface of the dots. The linear and nonlinear optical properties of small and large quantum dots are discussed in detail, including transient optical nonlinearities (photon echo) and two-photon absorption studies. The modifications of the electron-hole states introduced by spin-orbit coupling, carrier-lattice coupling (phonons), as well as static electric or magnetic fields are outlined.

Even though the main chapters of this book are based on work in which the authors were more or less directly involved, we try to give extensive references, also to papers which we did not use explicitly in our calculations. However, as a consequence of the overwhelmingly large number of publications in this field, we surely missed many valuable contributions of a large number of authors, and we are sorry for that.

We wish to thank H. Haug for many discussions, collaborations, and for encouraging and stimulating the Habilitation thesis of one of us (LB). This thesis is the basis of several chapters of this book. Furthermore, we wish to thank Y. Z. Hu and M. Lindberg for collaboration on many quantum-dot problems and Y. Z. Hu for the permission to use some figures from his dissertation. We thank E. L. Pollock for collaboration and helpful discussions of the quantum Monte Carlo method. We gratefully acknowledge stimulating quantum dot collaborations with our experimental colleagues, especially N. Peyghambarian, H. M. Gibbs, C. Klingshirn and their coworkers.

This manuscript was prepared using Scroll System's PS^{TM} Technical Word Processor. We acknowledge financial support from the DFG, NSF, OCC, ARO, AFOSR, the Volkswagen Foundation, and a NATO travel grant.

Frankfurt and Tucson   Ladislaus Bányai
March 1993   Stephan W. Koch

CONTENTS

SEMICONDUCTOR QUANTUM DOTS

Chapter 1
INTRODUCTION

The impressive progress in the fabrication of low-dimensional semi-conductor structures during the last decade made it possible to reduce the effective dimension from three dimensional bulk materials, to quasi-two dimensional quantum well systems, to quasi-one dimensional quantum wires, and even to quasi-zero dimensional quantum dots. The quantum confinement effects in such semiconductor systems with reduced space dimensions have attracted considerable attention. Especially interesting are the modified electronic and optical properties of these structures, which are controllable to a certain degree through the flexibility in the structure design. This feature makes quantum confined semiconductors very promising for possible device applications in microelectronics, nonlinear optics, and many other fields.

The best known examples of semiconductor systems with reduced space dimensions are the quantum-well structures which are made of alternating layers of active and transparent semiconductor materials. External excitation of these structures, e.g. by laser irradiation in the appropriate frequency regime, generates electron-hole pairs within the quasi-two-dimensional active layers. These layers provide confinement in one space dimension which is already sufficient to largely enhance excitonic effects.

It is well known, that in two dimensions the ground-state binding energy of Wannier excitons is four times larger than in the corresponding three-dimensional material. Therefore, exciton effects are easily observable in quasi-two-dimensional systems even at room temperature. Since these exciton resonances are strongly modified through the application of external fields or by the excitation of additional carriers in the system, the quantum-well systems show pronounced room-temperature optical nonlinearities.

Stimulated by these encouraging results in quantum wells, researchers have intensified the investigation of optical and electronic excitations of systems with even lower dimensionality. A large number of experimental and theoretical studies of the last years investigate the properties of excited carriers in the completely confined semiconductor quantum dot systems. These activities are motivated to a certain degree by the device application capability of these structures, but also pure scientific interest in the modified physical properties of these mesoscopic systems continues to stimulate work in this area.

In this book we review mostly the theoretical developments in the field of semiconductor quantum dot research. At the same time we make an attempt to include also some of the latest experimental developments, in particular where direct comparison of experimental and theoretical results are possible. The choice of topics covered in this book is strongly influenced by our own views and interests and we do not claim to present a complete discussion of the field.

Some of the first quantum dot systems were probably made many hundred years ago by people who created colored glass by melting a certain amount of semiconductor material, such as ZnS or ZnSe together with the usual glass material. The very dilute system of semiconductor crystallites in the glass absorbs light at the characteristic wavelengths, leading to a coloring of the glass. Of course, the size of these early nanocrystallites was not well controlled, leading to a substantial distribution of sizes and also material composition.

More controlled attempts to create quantum dots inside another matrix material started in the late seventies of this century and intensified during the eighties and early nineties. Originally, the majority of the investigations concentrated on quantum dots embedded in glass matrices or suspended in colloidal solutions. As it turns out, these crystallites have more or less spherical shapes and may be fabricated with very small radii (between 1 and 100 nanometers, nm).

The growth of semiconductor quantum dots in glasses is still one of the two most-used techniques [Ekimov et al. (1981), (1982), (1985); Warnock and Awschalom (1985), (1986); Borelli et al. (1987); Liu and Risbud (1990); Woggon et al. (1991)]. The samples are typically prepared by high-temperature precipitation in molten silicate glass matrices. After melting, the glasses are doped with the desired semiconductor components. A rapid cooling process of the melt is followed by a secondary heat treatment, i.e. tempering between 400 and 700 degrees C, during which the semiconductor crystallites precipitate out of the solid solution. The average radius of the crystallites at a given concentration of the semiconductor components in the glass matrix (supersaturation) is determined by the temperature and duration of the secondary heat treatment. Lifshitz and Slezov (1959) showed that under idealized conditions the average cluster radius follows a simple growth law and the radius distribution assumes a universal shape (see Appendix). Even though some indications of a growth law for the average cluster radius have been reported [Ekimov et al. (1985)], no definite experimental verification of a universal size distribution has been reported yet. The uncertainty of the dot size distribution leads to some complication in the interpretation of optical experiments, since the size dispersion causes pronounced inhomogeneous broadening of the resonances. These features will be discussed in more detail in Sec. 2-3 of this book. To overcome this problem, sophisticated experiments and sample preparation techniques are currently being developed to study optical properties of single quantum dots [Brunner et al. (1992)].

Generally, even identical semiconductor quantum dots inside a matrix with different dielectric properties represent an optically inhomogeneous system. The task of determining the optical properties of the mixture, i.e. quantum dots plus matrix, from those of its components [Garnett (1904)] is sometimes complicated and still the subject of current investigations [Sheu et al. (1990)]. Fortunately, in many cases the concentration of the dots is sufficiently low, so that a simplified linear interpolation formula weighted only with the respective concentration of dot and matrix material might be applied. In most parts of this book we discuss therefore the properties of single noninteracting quantum dots.

Besides the growth of quantum dots inside a glass matrix, the synthesis of semiconductor crystallites in liquid solvents is the second one of the currently most used techniques [see e.g. Brus (1984), (1986), Weller et al. (1986), Alivisatos et al. (1988)]. The procedures applied are typically based on near-room-temperature precipitation using methods and materials borrowed from organometallic and polymer chemistry. This technique has basically the same advantages and problems as the growth of quantum dots in glass, but it offers the possibility of manipulating the clusters after the original growth. Surface passivated "capped" clusters have been fabricated by growing a wider bandgap material around the original cluster or by bonding phenyl groups to the crystallites [Rossetti et al. (1984), Brus (1991)]. Even though these procedures are very promising, the materials produced are in many respects not superior to the semiconductor doped glasses.

Semiconductor microcrystallites have been grown in a crystal matrix [Tsuboi (1980)], inside ionic materials [Masumoto et al. (1989)], in polymer films [Hilinski et al. (1988)], or even inside the cavities of certain zeolites [Wang and Herron (1987), Wang et al. (1989), Grätzel, (1989), Nozue et al. (1990), Tang et al. (1991)]. Even biosynthesis of quantum dots has been reported in certain yeasts [Dameron et al. (1989)]. Quantum-confinement-induced absorption shifts have been observed in these materials, but generally no strongly pronounced optical resonances have been reported. Sometimes modifications of the quantum dot structure occur because the semiconductor material interacts with the host material, leading to structure changes and additional broadening mechanisms.

Molecular precursor pyrolysis reactions are a relatively recent synthesis technique which seems promising for the production of III-V semiconductor quantum dots [Olshavsky *et al*. (1990), Douglas and Theopold (1991), Uchida *et al*. (1992)]. For example, GaAs clusters have been prepared which show some confinement-induced absorption shift, but so far without good resonance structure.

In another recent development, quantum dots have been grown in matrices which are synthesized by the sol-gel technique [Bagnall and Zarzycki (1990), MacKenzie (1992)]. By using porous wet gel as a matrix into which reactants could be diffused and subsequently precipitated, the size of the pores and their distribution through the gel controls both the average size and the distribution of the quantum dots. Relatively high concentrations of dots have been produced this way, but the measured linewidth of the absorption profile is still rather large.

Another very active area is the preparation of quantum dots through etching or ion-implantation of quasi-two-dimensional semiconductor quantum wells [see e.g. Reed *et al*. (1986), (1988), Kash *et al*. (1986), Cibert *et al*. (1986), Temkin *et al*., (1987), Hansen *et al*. (1988), Zarem *et al*. (1989), Tsuchiya *et al*. (1989), Demel *et al*. (1990), Vahala (1990)]. Mostly, quantum dots in AlGaAs based material have been obtained with radii typically in the range of 50 nm or larger. Using similar techniques, quantum dots of InSb [Sikorski and Merkt (1989), Merkt (1990)] and GaInAs/InP [Galeuchet *et al*. (1991)] have been produced. Effective three-dimensional quantum confinement has also been obtained through application of mechanical strain [Kash *et al*. (1989), (1990), Kash (1990)] and, for electrons only, through field effects [Alsmeier *et al*. (1990), Kotthaus *et al*. (1990)]. Recent summaries of the developments in the field of etched quantum dots can be found in the reviews by Kastner (1993) and Reed (1993).

Until recently, mostly direct-gap semiconductor materials have been used in the preparation of quantum dots. Most typical are here the III-V and II-IV compounds, but also CuCl and CuBr crystallites are being studied extensively. The general theoretical representation of the electronic states in these quantum dots consists primarily in the extension of concepts which already have been applied successfully for quantum well structures. Basically, one assumes that the difference of the band gap of the spatially confined semiconductor to that of its surrounding (another semiconductor or insulator) acts as a potential barrier for the conduction-band electrons and the valence-band holes. In many practically realized systems this barrier is very high and allows, as a first approximation, an idealization through infinite walls. In addition, however, there exist also very interesting quantum dot systems where the confinement barrier is relatively low, or where one has a smooth transition from the dot to the barrier material (graded barriers). Nevertheless, even for a relatively high barrier, such as for quantum dots in a glass matrix, the structure of the surrounding material is not always without relevance. For example, inside the spectrally wide transparency region of glasses, typically one still has a large number of localized states due either to glassy disorder or doping. These states might be accessible through tunneling out of the quantum dot [Roussignol *et al.* (1987)].

Even though the vast majority of quantum dot systems is based on direct-gap semiconductors, recently also indirect-gap quantum dots have been realized [Brus (1992), Maeda (1991)]. It has been found that the luminescence intensity of these quantum dots is enhanced by many orders of magnitude over the indirect-gap bulk material. Even though the recent developments of indirect-gap quantum dots are extremely interesting [Takagahra and Takeda (1992)], in this book we will mostly focus on direct-gap materials, since these systems are currently much better understood.

The spatial dimensions of the quantum dots are usually big in comparison to the lattice constant of the semiconductor material. Therefore, it is a reasonable first assumption to consider the band structure of these mesoscopic systems as only weakly changed in comparison to the corresponding bulk material. This approach is often called *envelope function approximation* since one assumes that only the envelope part of the wavefunction is modified through the existence of the confinement potential. For the states in the vicinity of the fundamental absorption edge, one often uses the effective mass approximation for the envelope function. This means, that instead of plane waves, as in bulk materials, the envelope function is the wavefunction of a particle in the appropriate three-dimensional potential well.

Through laser excitation of quantum dots with energies in excess of the bandgap energy, electrons in the conduction band and simultaneously holes in the valence band are excited. These charged particles interact through Coulomb forces which are screened by the core electrons and ions. We will show that in this context it is important also to take into account the effects of surface polarization, which occur due to the different dielectric conditions in the material of the dot and its surrounding medium.

Depending on the excitation conditions the Coulomb attraction between an electron and a hole in a bulk semiconductor might lead to a bound state, the Wannier exciton. Just as the hydrogen atom, the exciton states are characterized by a product wavefunction consisting of a plane-wave part for the center-of-mass motion and hydrogen functions for the electron-hole relative motion. The characteristic length scale for the relative motion is the exciton Bohr radius, which may be of the order of 1 - 20 nm, depending on the semiconductor material. Quantum confinement effects arise, as soon as the linear extension of the quantum dot is comparable to this exciton Bohr radius, leading to confinement induced modifications of the electron-hole pair states. Such semiconductor nanostructures, whose dimensions are large in comparison to the lattice constant but comparable to the exciton Bohr radius, are called *mesoscopic structures*.

A substantial fraction of the experimental and theoretical investigations of quantum dot structures are devoted to the interband optics in the exciton range. The basic ideas for the theoretical treatment of the Coulomb interaction in this domain have been developed by Efros and Efros (1982) and Brus (1984). Depending on the magnitude of the crystallites one generally differentiates between different regimes of confinement-quantization and applies specific approximations.

The existence of an extensive literature dealing with two-dimensional quantum-well structures allows us to concentrate only on those aspects, which are specific for quantum dots, i.e. which result from the three-dimensional quantum confinement. For the general theoretical problem dealing with the contact between two semiconductor layers we refer the interested reader to the book by Bastard (1988). An extensive theoretical discussion of the many-body aspects of optical nonlinearities may be found in the modern books of Haug and Koch (1990, 1993) and Zimmermann (1988). Good collections of contributions dealing with the properties of confined semiconductors including quantum dots are also available [Haug and Banyai (1989), Kuchar *et al.* (1990), D'Andrea *et al.* (1992), special issue of Physica (March 1993)].

Chapter 2
THEORETICAL CONCEPTS

In this chapter we provide some background material and theoretical concepts which are needed in the later parts of this book. First, we briefly summarize some relevant results from elementary quantum mechanics dealing with the quantized motion of a particle in a potential well (Sec. 2-1). In Sec. 2-2 we then discuss the basic properties of optical transitions in quantum dots, assuming the idealized case of a two-band semiconductor with parabolic bands. For the conditions of perfect quantum confinement, i.e. infinitely high potential barriers for the carriers in the dot, we compute the excitation spectrum of the system in the electron-hole representation where electrons and holes are characterized by simple scalar effective masses.

Most realistic quantum dot systems contain dots of various radii making it necessary to include the dot size distribution. Since the optical resonance energies strongly depend on the quantum dot radius, a radius distribution leads to a resonance distribution which manifests itself as in-homogeneous broadening in the optical spectra. These features are analyzed in Sec. 2-3 of this chapter.

In order to use theoretical results obtained for the susceptibility of a single quantum dot to compute the effective susceptibility of an experimental system containing a certain concentration of randomly distributed dots one can apply the Maxwell-Garnett theory. This theory, together with a basic analysis of local field effects is described in Sec. 2-4.

Typically, a quantum dot is surrounded by another dielectric medium. Hence, the optically generated charge carriers inside the dot induce image charges in the surrounding material, i.e., they lead to a surface polarization of the dot. In Sec. 2-5 we discuss how these effects can be incorporated into the Coulomb interaction between the carriers inside the dot. For this purpose we analyze the electrostatic properties of the quantum dot in its surrounding material.

2-1. Particle in a Potential Well

We begin our analysis of quantum confinement effects by considering the stationary Schrödinger equation for the motion of a single particle in a square potential well,

$$\left\{ -\frac{\hbar^2}{2m} \nabla^2 + U(\mathbf{r}) \right\} \psi(\mathbf{r}) = E \, \psi(\mathbf{r}) \ . \tag{2.1}$$

The potential energy is assumed to vanish inside a finite domain and to take a positive value U_0 outside that domain. The simplest problem of this kind is the one-dimensional system which allows particle motion along the x-axis in the potential

$$U(x) = \begin{cases} 0 & \text{for} \ |x| < a \\ U_0 & \text{for} \ |x| > a \end{cases}, \tag{2.2}$$

where we assume $U_0 > 0$. The potential $U(x)$ is schematically shown in Fig. 2.1.

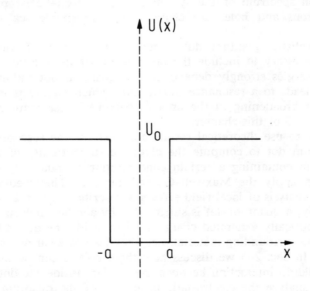

Fig. 2.1: The one-dimensional potential well.

From elementary quantum mechanics we know that the Schrödinger equation (2.1) yields even and odd eigenstates. Properly normalized, the wavefunctions of these states can be written as

$$\psi_k^{(+)}(x) = \frac{1}{\sqrt{a+\dfrac{1}{\kappa}}} \begin{cases} \cos(kx) & \text{for } x < a \\ \cos(ka)\, e^{\kappa(a\,-\,x)} & \text{for } x > a \end{cases} \qquad (2.3)$$

for the even solutions and

$$\psi_k^{(-)}(x) = \frac{1}{\sqrt{a+\dfrac{1}{\kappa}}} \begin{cases} \sin(kx) & \text{for } x < a \\ \sin(ka)\, ed\kappa(a\,-\,x) & \text{for } x > a \end{cases} \qquad (2.4)$$

for the odd solutions, where even and odd is defined with respect to reflections around the origin. The one-dimensional bound-state energies

$$E_k = \frac{\hbar^2}{2m} k^2 < U_0 \qquad (2.5)$$

are given by the solutions of the transcendental equation

$$tg(ka) = \frac{\sqrt{k_0^2 - k^2}}{k} \qquad (2.6)$$

for the even states and

$$ctg\left(\frac{ka}{2}\right) = -\frac{\sqrt{k_0^2 - k^2}}{k} \qquad (2.7)$$

for the odd states, respectively. The parameters k_0 and κ are defined through

$$U_0 = \frac{\hbar^2}{2m} k_0^2 \qquad (2.8)$$

and

$$\kappa^2 = k_0^2 - k^2 .$$

(2.9)

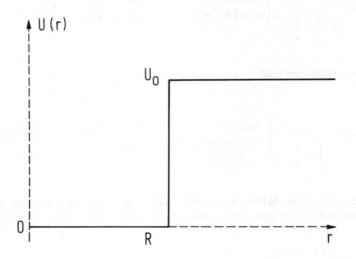

Fig. 2.2: The spherical potential well.

For infinitely deep high potential barriers ($U_0 \to \infty$) the wavefunctions vanish outside the well. Inside the well the wavefunctions are

$$\psi_k^{(+)}(x) = \frac{1}{\sqrt{a}} \cos(kx)$$

(2.10)

for the even states and

$$\psi_k^{(-)}(x) = \frac{1}{\sqrt{a}} \sin(kx)$$

(2.11)

for the odd states, respectively. The corresponding energy eigenvalues are

$$k = \frac{(2n + 1)\pi}{2a} \quad (n = 0, 1, 2, ...)$$

(2.12)

and

$$k = \frac{n\pi}{a} \quad (n = 1, 2, 3, ...) \ . \tag{2.13}$$

For the infinite potential barrier ($U_0 = \infty$) one can construct the solution for the three-dimensional problem as product of the wavefunctions of the one-dimensional well, Eqs. (2.10) - (2.11). This procedure, however, fails for potential wells of finite depth. The only three-dimensional potential with finite depth ($U_0 < \infty$), for which a separation of variables is possible, is the spherical potential

$$U(r) = \begin{cases} 0 & \text{for } r < R \\ U_0 & \text{for } r > R \end{cases}, \tag{2.14}$$

which is plotted in Fig. 2.2.

Considering the angular momentum conservation one can reduce the three-dimensional problem to an one-dimensional eigenvalue problem. The eigenfunctions can be written as

$$\psi_{\ell,m,k}(\mathbf{r}) = \phi_{\ell,m,n}(r) \, Y_{\ell m}(\theta,\phi) \ , \tag{2.15}$$

where $Y_{\ell m}(\theta,\phi)$ are the spherical harmonics, and the quantum numbers are restricted to the values

$n = 1, 2, 3, ...$

$\ell = 0, 1, 2, ...$

$m = 0, \pm 1, \pm 2, ... , \pm \ell \ .$

The radial part of the wavefunction satisfies the radial Schrödinger equation

$$-\frac{1}{r} \frac{d^2}{dr^2} \left[r \, \phi_{\ell,m,n}(r) \right] + \left(\frac{\ell(\ell+1)}{r^2} - k^2 \right) \phi_{\ell,m,n}(r) = 0 \tag{2.16a}$$

for $r < R$, and

$$-\frac{1}{r} \frac{d^2}{dr^2} \left[r \, \phi_{\ell,m,n}(r) \right] + \left(\frac{\ell(\ell+1)}{r^2} + \kappa^2 \right) \phi_{\ell,m,n}(r) = 0 \tag{2.16b}$$

for $r > R$. The solutions which are regular in the origin and at infinity are given by the spherical Bessel functions $j_\ell(kr)$ and $h_\ell(i\kappa r)$, respectively [Abramowitz and Stegun (1970)]. The energy eigenvalues, or allowed k - values, are determined by the continuity of the logarithmic derivatives of these functions. A typical feature of the three-dimensional well is that no bound states exist in the well for $k_0 < 1/2$, i.e., $U_0 R^2 < \hbar^2/8m$.

For later reference, we list in the following the explicit form of the normalized wavefunctions for the case of an infinite potential well, which always has infinitely many bound states. The wavefunctions are

$$\phi_{\ell,m,k}(r) = \sqrt{\frac{2}{R^3}} \frac{j_\ell(kr)}{j_{\ell+1}(kR)} , \tag{2.17}$$

and the energy eigenvalues are determined from the zeros of the spherical Bessel functions of the first kind,

$$j_\ell(\kappa_{n,\ell}) = 0 \quad , \tag{2.18}$$

and $k = \kappa_{n,\ell}/R$.

Also for later reference we summarize here some important properties of the spherical Bessel functions for $\ell = 0$ and $\ell = 1$. The $\ell = 0$ Bessel function is

$$j_0(r) = \frac{\sin(r)}{r} \tag{2.19}$$

with the roots

$$\kappa_{n,0} = n\pi \quad (n = 1, 2, 3, \dots) . \tag{2.20}$$

The $\ell = 1$ Bessel function is

$$j_1(r) = \frac{\sin(r)}{r^2} - \frac{\cos(r)}{r} \tag{2.21}$$

and the roots are solutions of the transcendental equation

$$tg(\kappa_{n,1}) = \kappa_{n,1} . \tag{2.22}$$

The first solution of Eq. (2.22) is $\kappa_{1,1} = 4.493$, i.e. a lower value than the second root for $\ell = 0$ which we obtain from Eq. (2.20) as $\kappa_{2,0} = 2\pi$.

Generally, the asymptotic behavior of the spherical Bessel functions is given by

$$j_\ell(r) \simeq \frac{\sin(r - \ell\pi/2)}{r} \ , \qquad (r \to \infty) . \tag{2.23}$$

Consequently, the asymptotic roots of j_ℓ for fixed ℓ are equidistant multiples of π.

In view of the current development of quasi-two-dimensional quantum dots, we give here also solutions for a truly two-dimensional quantum disk. The two-dimensional Laplace operator is

$$\Delta = \frac{\partial^2}{\partial r^2} + \frac{1}{r} \frac{\partial}{\partial r} + \frac{1}{r^2} \frac{\partial^2}{\partial \phi^2} . \tag{2.24}$$

The eigenfunctions of the kinetic energy operator are eigenstates of the angular momentum component along the z-axis perpendicular to the disk,

$$\psi(r,\phi) = \psi(r) \, e^{iM\phi} . \tag{2.25}$$

The radial wavefunctions satisfy the Schrödinger equation

$$- \frac{\hbar^2}{2m} \left[\frac{\partial^2}{\partial r^2} + \frac{1}{r} \frac{\partial}{\partial r} - \frac{M^2}{r^2} + E \right] \psi(r) = 0 \tag{2.26}$$

with the proper solutions given by the Bessel functions

$$J_M(\lambda r) = \frac{1}{\pi} \int_0^\pi d\theta \, \cos(M\theta - \lambda r \sin\theta) \tag{2.27}$$

with $\lambda = (2mE/\hbar^2)^{1/2}$. For an infinite potential barrier the energy eigenvalues are given by the zeros of the Bessel function (2.27).

2-2. Optical Transitions

To discuss the basic properties of optical transitions in quantum dots we assume in this section a simple two-band semiconductor with parabolic, isotropic, direct bands which are only spin degenerate. We use the effective mass approximation for the carriers in the quantum dot. These approximations imply that the dots have a linear dimension which substantially exceeds the lattice constant of the bulk semiconductor. To describe the excitation spectrum of the system we use the electron–hole representation where electrons and holes have the effective masses m_e and m_h, respectively. As further idealization, we approximate the difference of the band gaps of the semiconductor dot and of the surrounding material as an infinitely high potential barrier. Consequently, the motion of the electrons and holes is confined inside the dot.

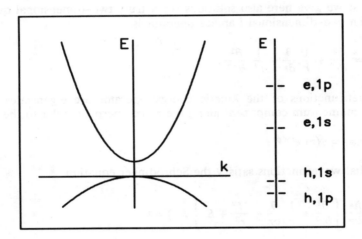

Fig. 2.3: Schematic plot of the single particle energy spectrum in bulk semiconductors (left). The single particle energies for electrons (e) and holes (h) in small quantum dots are shown in the right part of the figure.

We discuss here specifically the ideal case of spherical quantum dots with a radius R. For the spherical potential well we have shown in the previous section that the kinetic energy of a quantum mechanical particle takes discrete values and scales with the square of the inverse radius,

$$E_{e,n\ell m} = E_g + \frac{\hbar^2}{2m_e}\left(\frac{\kappa_{n,\ell}}{R}\right)^2 \quad \text{and} \quad E_{h,n\ell m} = \frac{\hbar^2}{2m_h}\left(\frac{\kappa_{n,\ell}}{R}\right)^2, \qquad (2.28)$$

where E_g is the bandgap energy of the respective bulk semiconductor. It is usual to refer to the n, ℓ electron or hole eigenstate as $1s$, $1p$, $1d$, etc. , where s, p, d, etc., correspond to $\ell = 0$, 1, 2, ..., respectively. Note, that a $1p$ state in a spherical confinement potential is possible in contrast to the well-known case of a Coulomb potential. The first few of the energy states (2.28) are shown schematically in Fig. 2.3 in comparison to the continuous energy dispersion of the bulk semiconductor material.

From Eq. (2.28) we see that the lowest confined state of a single electron-hole pair has its energy increased with respect to the bulk semiconductor band gap by

$$\Delta E = \frac{\hbar^2}{2m_r} \left[\frac{\pi}{R} \right]^2 , \qquad (2.29)$$

where m_r is the reduced electron-hole mass

$$\frac{1}{m_r} = \frac{1}{m_e} + \frac{1}{m_h} . \qquad (2.30)$$

In order to introduce scaled quantities, we use the exciton Rydberg energy

$$E_R = \frac{\hbar^2}{2m_r a_B^2} , \qquad (2.31)$$

and the exciton Bohr radius

$$a_B = \frac{\epsilon_1 \hbar^2}{m_r e^2} , \qquad (2.32)$$

where we always refer to the bulk semiconductor material with the (background) dielectric constant ϵ_1. In terms of these quantities the energy shift (2.29) can be written as

$$\Delta E = E_R \left[\frac{\pi a_B}{R} \right]^2 . \qquad (2.33)$$

This equation shows that for small quantum dots ($R \ll a_B$) the confinement induced energy shift is large in comparison to the exciton binding energy. If we use the exciton binding energy as a measure of the importance of interband Coulomb effects, we can argue that the large energy shifts for small dots suggests to ignore Coulomb interactions in small dots. We will show in later chapters that this is a dangerous approximation which ignores some of the important physics in semiconductor quantum dots. However, this approximation allows to obtain some important insight and has the major advantage of yielding analytic solutions.

In a first approximation we therefore apply a simple model of non-interacting particles [Efros and Efros (1982), Brus (1984) and (1986), Schmitt-Rink *et al.* (1987)]. The Hamiltonian of such a non-interacting electron-hole system is

$$\mathcal{H} = \sum_{\alpha = e, h} \int d^3r \, \psi_\alpha^\dagger(\mathbf{r}) \left(-\frac{\hbar^2 \nabla^2}{2m_\alpha} + \frac{E_g}{2} \right) \psi_\alpha(\mathbf{r}) \ , \qquad (2.34)$$

where the field operators ψ_α satisfy the Fermi anti-commutation relations

$$[\psi_\alpha^\dagger(\mathbf{r}) , \psi_\beta(\mathbf{r}')]_+ = \psi_\alpha^\dagger(\mathbf{r}) \, \psi_\beta(\mathbf{r}') + \psi_\beta(\mathbf{r}') \, \psi_\alpha^\dagger(\mathbf{r})$$

$$= \delta_{\alpha,\beta} \, \delta(\mathbf{r} - \mathbf{r}') ; \quad (\alpha,\beta = e, h) \ , \qquad (2.35a)$$

$$[\psi_\alpha(\mathbf{r}) , \psi_\beta(\mathbf{r}')]_+ = 0 \ , \qquad (2.35b)$$

and

$$[\psi_\alpha^\dagger(\mathbf{r}) , \psi_\beta^\dagger(\mathbf{r}')]_+ = 0 \ . \qquad (2.35c)$$

The boundary condition is

$$\psi_\alpha(\mathbf{r}) \Big|_{r = R} = 0 \ . \qquad (2.36)$$

The field operators can be expanded in terms of the eigenfunctions (2.17),

$$\psi_\alpha(\mathbf{r}) = \sum_{\ell m n} a_{(\alpha)\ell m n} \, \phi_{\ell,m,k}(\mathbf{r}) \ . \qquad (2.37)$$

Here we introduced the particle creation and annihilation operators $a_\alpha^\dagger$ and a_α, $\alpha = e, h$, of electrons or holes in the respective states. The electron and hole operators fulfill the Fermi anti-commutation relations in the form

$$[a_{(\alpha)\ell mn}, a_{(\alpha')\ell'm'n'}^\dagger]_+ = \delta_{\alpha,\alpha'}\,\delta_{\ell,\ell'}\,\delta_{m,m'}\,\delta_{n,n'}\,, \tag{2.38a}$$

$$[a_{(\alpha)\ell mn}, a_{(\alpha')\ell'm'n'}]_+ = 0\,, \tag{2.38b}$$

and

$$[a_{(\alpha)\ell mn}^\dagger, a_{(\alpha')\ell'm'n'}^\dagger]_+ = 0\,. \tag{2.38c}$$

The Hamiltonian (2.34) can be written in terms of the electron and hole operators as

$$\mathcal{H} = \sum_{\alpha \ell mn}\left[\frac{E_g}{2} + \frac{\hbar^2}{2m_\alpha}\left(\frac{\kappa_{n,\ell}}{R}\right)^2\right] a_{(\alpha)\ell mn}^\dagger\, a_{(\alpha)\ell mn}\,. \tag{2.39}$$

Other important operators which we will encounter in the discussion of the following chapters include the number operator for electrons (holes)

$$N_\alpha = \int d^3r\,\psi_\alpha^\dagger(\mathbf{r})\,\psi_\alpha(\mathbf{r})\,, \tag{2.40}$$

the charge density operator

$$\rho(\mathbf{r}) = e\,[\,\psi_h^\dagger(\mathbf{r})\,\psi_h(\mathbf{r}) - \psi_e^\dagger(\mathbf{r})\,\psi_e(\mathbf{r})\,]\,, \tag{2.41}$$

where e is the absolute value of the electron charge, and the interband polarization operator

$$P = p_{cv}\int d^3r\,\psi_e(\mathbf{r})\,\psi_h(\mathbf{r}) + \text{h.c.} = p_{cv}\sum_{\ell mn} a_{(e)\ell mn}\, a_{(h)\ell\text{-}mn} + \text{h.c.}\,, \tag{2.42}$$

where p_{cv} is the polarization matrix element between conduction and valence band.

The coupling of the interband polarization to the optical electric field via the interaction Hamiltonian,

$$\mathcal{H}_I \propto - P E \cos(\omega t) , \tag{2.43}$$

describes the optical interband transitions. Writing the interband polarization operator in the form of Eq. (2.42) follows the usual notion to attribute the constant dipole matrix element p_{cv} between the conduction and the valence bands to the Bloch part of the wavefunctions. Furthermore, we assumed that the electric field is oriented along the axis of the interband dipole moment. More details of the dipole coupling will be discussed in Sec. 5-1.

From Eq. (2.42) it follows that optical interband transitions involve only electrons and holes with identical orbital angular momentum ℓ and opposite azimuthal quantum number m. Furthermore, the radial quantum numbers n of the dipole-coupled electrons and holes are also identical. This, as well the independence of the oscillator strengths from the quantum numbers of the created pair, is actually a direct consequence of the identity of the spatial electron and hole wavefunctions, i.e. of the independence of the wavefunctions on the mass of the particle. Another consequence of the same fact is that optically created neutral electron-hole systems have a vanishing local charge density. We will show in later chapters, especially in Chap.4, that this idealized assumption has to be modified in a more realistic analysis which includes electron-hole Coulomb interaction and spin-orbit coupling.

Although the wavefunctions of electrons and holes in the noninteracting effective mass theory have the same form, their energies are different due to the different electron and hole masses. The photon energy needed for the creation of an electron-hole pair $(n,\ell,\pm m)$ is

$$\hbar\omega = E_g + E_R \left[\frac{\kappa_{n,\ell} \, a_B}{R} \right]^2 . \tag{2.44}$$

A simple calculation, the details of which are given in Sec. 5-2, yields the optical interband susceptibility of a single quantum dot in our current model as

$$\chi(\omega) = -\frac{|p_{cv}|^2}{\Omega} \sum_{n,\ell} \frac{2\ell+1}{\hbar\omega - E_g + E_R \left[\dfrac{\kappa_{n,\ell} a_B}{R}\right]^2 + i\gamma} + (\omega \rightarrow -\omega) ,$$

(2.45)

where $\Omega = 4\pi R^3/3$ is the dot volume and γ is a phenomenological linewidth (dephasing rate). The factor $2\ell + 1$ results from the m degeneracy of the levels.

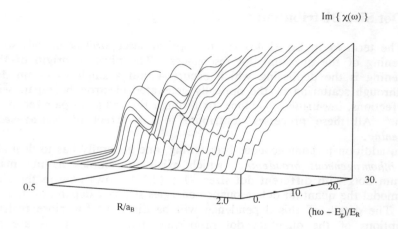

Im $\{ \chi(\omega) \}$

Fig. 2.4: Imaginary part of the quantum dot susceptibility, Eq. (2.45). The homogeneous linewidth was chosen as $\gamma = 0.2\ E_R$. Furthermore, we included an averaging over the quantum dot radius, Eq. (2.48), using the Lifshitz-Slezov distribution (see Sec. 2-3 and Appendix). This figure is also printed on the cover of this book.

Comparing the oscillator strength of Eq. (2.45) with the well known result for excitons in bulk semiconductors,

$$\frac{|p_{cv}|^2}{\pi a_B^3} ,$$

we note an amplification of the quantum-dot oscillator strength by the factor $(a_B/R)^3$. This enhancement influences both the real and imaginary parts of the susceptibility, so that it is usually not very important for applications, where typically the ratio of real and imaginary part of the susceptibility has to be considered [Peyghambarian and Koch (1990)].

In Fig. 2.4 we plot the imaginary (absorptive) part of the susceptibility, Eq. (2.45), as a function of frequency and dot radius. The results clearly illustrate the absorption enhancement as well as the blue shift of the energy levels with decreasing radius.

2-3. Dot Size Distribution

The term $i\gamma$ in Eq. (2.45) for the optical susceptibility models some broadening of the quantum dot resonances. The physical origin of this broadening is the dephasing which occurs within a single quantum dot, e.g., through scattering of the optically generated electron-hole-pair with imperfections, impurities, phonons, or through the radiative pair recombination. All these processes contribute to the effect of *homogeneous broadening*.

In addition to homogeneous broadening, one generally has to deal also with *inhomogeneous broadening*, which exists in samples having many quantum dots with different dot sizes. Eq. (2.28) shows how in the simplest model the quantum dot resonance energies strongly depend on the dot size. The details of this dependence will be different for more realistic descriptions of the quantum dot properties, but these aspects are not important for our current analysis. All we need to realize here is that a certain distribution of dot sizes automatically causes a distribution of the resonance energies. Each quantum dot size contributes to the total absorption with a weight given by the probability to find this particular size in the sample. If we denote by $\alpha(\omega)|_R$ the absorption coefficient for a dot of radius R we can compute the average dot absorption coefficient

$$\alpha(\omega)\big|_{av} = \int_0^\infty dR\ P(R)\ \alpha(\omega)\big|_R\ , \qquad (2.48)$$

where $P(R)$ is the probability distribution of dot radii [Wu *et al.* (1987)], see also the following Sec. 2-4.

To illustrate the effect of inhomogeneous broadening for a non-universal distribution, let us assume that the sizes are statistically distributed around a mean radius $\overline{R}$. Using for $P(R)$ a Gaussian distribution gives the results shown in Fig. 2.5 This figure illustrates how the energetically lowest one-pair resonances are broadened with increasing width of the dot-size distribution function. The originally discrete resonances merge into a continuous structure for widths around 10%. Spectra such as those for 15 - 20 % broadening are typical for experimentally realized semiconductor doped glasses or colloids.

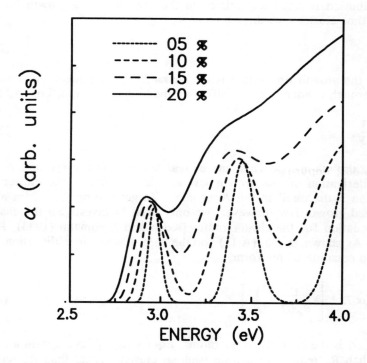

Fig. 2.5: Linear absorption for quantum dots with a Gaussian size distribution. The different curves are for the widths of the Gaussian size distribution indicated in the figure.

Generally, a Gaussian size distribution is a reasonable assumption if one has no further information about the quantum dot sample. There are, however, situations where additional information is available. For example, a quantum dot may grow by a well-defined process, such as growth by condensation of monomers which diffuse from the surrounding medium to the cluster surface. For this case, which may be realized for quantum dots in a glass matrix, one may have a universal dot size distribution which is only determined by the growth parameters of the system. The computation of such a universal size distribution requires the asymptotic analysis of the governing kinetic equations of the growing clusters in the supersaturated surrounding. Following the classical Lifshitz - Slezov (1961) work we summarize this procedure in the Appendix. Here we discuss only the main results.

In the Appendix we show that the existence of an asymptotic cluster size distribution is intimately related to the existence of a growth law for the long-time cluster growth,

$$\overline{R} \propto t^\delta . \tag{2.49}$$

Here t is the growth time and δ is the characteristic exponent. For cluster growth through condensation of diffusing monomers we find, Eq. (A.65),

$$\delta = \frac{1}{3} . \tag{2.50}$$

However, also other exponents may be realized in different systems.

The derivation of the growth law and its relation to the cluster size distribution is discussed in the Appendix. To get a superficial idea about the realized growth law, however, we only need to investigate the characteristic equation for the cluster radius [Koch and Liebmann (1983), Koch (1984)]. As shown in Eq. (A.15) for the case of monomer diffusion, one obtains an equation of the form,

$$\frac{dR}{dt} \propto R^{-\lambda} \left[\frac{1}{R_c(t)} - \frac{1}{R} \right] , \tag{2.51}$$

where $R_c(t)$ is the critical cluster radius, Eq. (A.18). Clusters with a radius smaller than R_c tend to evaporate whereas clusters larger than R_c tend to grow. This critical radius is basically determined by the degree of supersaturation in the medium which surrounds the cluster. If the supersaturation is high, even small clusters can grow and R_c is small, whereas a low supersaturation leads to a large value of R_c and the evaporation of smaller clusters (see Appendix for more details).

The exponent λ in Eq. (2.51) is the main characteristics of the growth mechanism. In Eq. (A.15), we find $\lambda = 1$ for the case of monomer diffusion, whereas for cluster growth through condensation of kinematically moving monomers we have $\lambda = 0$ [Desai et al. (1983), Koch (1984)].

An equation such as (2.51) can only have a power-law solution of the form (2.49) if

$$R_c(t) \propto R(t), \tag{2.52}$$

and

$$\delta = (\lambda + 2)^{-1}. \tag{2.53}$$

Eq. (2.52) is indeed found in the full asymptotic analysis, see Eq. (A.64), and Eq. (2.53) yields the result (2.50) for the case of diffusing monomers, and $\delta = 1/2$ for kinematically moving monomers, respectively.

Fig. 2.4 shows an example of computed quantum-dot resonances for different dot sizes using the Lifshitz-Slezov distribution.

2-4. Susceptibility of Quantum Dot Systems

Unlike an idealized bulk semiconductor the system of quantum dots inside a matrix material is structurally inhomogeneous on the mesoscopic scale even though it is more or less translationally invariant on the macroscopic scale. To incorporate these features in the theoretical analysis one has to adopt an electrodynamic approach which takes into account wave propagation aspects of the light. Generally, such a theory is quite complicated and requires knowledge of the spatial distribution of the composite material. Fortunately, however, for most realistic systems such a wave propagation analysis is not really necessary since typically the wavelength λ of the light used to excite the band-edge states of quantum dots is much bigger than the dot radius.

Let us illustrate this point with some numbers using the exciton Bohr radius as the reference for the quantum dot sizes. The Bohr radius are approximately 140 Å in GaAs and 12 Å in CuCl. On the other hand the semiconductor bandgaps vary between approximately 1.5 eV in GaAs and 3.2 eV by CuCl, which corresponds to wavelengths of 7646 Å and 3680 Å in vacuum. The refractive index of these materials reduces these wavelengths approximately by a factor of three. Therefore, the condition of $R \ll \lambda$ is basically always well satisfied.

It can be shown [Sommerfeld (1949), Bottcher (1973)] that for $R \ll \lambda$ an analysis of static (non-propagating) electric fields reveals relationships which are also correct for propagating waves. The only modification is that the static dielectric constants are replaced by the respective complex, frequency dependent dielectric functions. In what follows we therefore present the simple static analysis to obtain the local field effects in quantum dots and to derive the so-called Maxwell-Garnett formula [Garnett, (1904)] for the effective susceptibility of a system of microspheres embedded in a dielectric medium.

Let us consider first a single dielectric sphere of radius R and dielectric constant ϵ_1 which is embedded in an infinite medium with the dielectric constant ϵ_2. We compute the potential $V(\mathbf{r})$ for a homogeneous, static field $\mathbf{E}$ which is oriented along the z-axis [Sommerfeld (1949), Bottcher (1973)]. The potential inside and outside the sphere must satisfy the Laplace equation (no sources)

$$\nabla^2 V(\mathbf{r}) = 0 . \qquad (2.54)$$

For infinite distance from the dot the potential has to behave like that of the homogeneous field,

$$V(\mathbf{r}) \propto -\mathbf{E}\cdot\mathbf{r} \quad (\text{for } r \to \infty) . \qquad (2.55)$$

At the boundary of the sphere the continuity of the tangential components of the field $\mathbf{E}$, respectively of the normal components of the induction field $\mathbf{D}$ must be assured. In spherical coordinates these boundary conditions imply the following conditions for the potential,

$$V\bigg|_{r=R-0} = V\bigg|_{r=R+0} \qquad (2.56)$$

$$\epsilon_1 \frac{\partial V}{\partial r}\bigg|_{r=R-0} = \epsilon_2 \frac{\partial V}{\partial r}\bigg|_{r=R+0} . \qquad (2.57)$$

We look for the solution inside the sphere as that corresponding to a homogeneous field $\mathbf{E}_i$ and outside the sphere as a superposition of the potential of the homogeneous applied field $\mathbf{E}$ and that of a yet undefined dipole d placed in the center of the sphere,

$$V(\mathbf{r}) = \begin{cases} - E_i \, r \cos\theta & \text{for } r < R \\ - E \, r \cos\theta - \dfrac{d}{r^2} \cos\theta & \text{for } r > R \end{cases} . \qquad (2.58)$$

Both expressions are solutions of the Laplace equation (2.54), but we still have to satisfy the boundary conditions, Eqs. (2.56) - (2.57). Due to the identical angular dependence of the ansatz (2.58) the boundary conditions reduce to the following conditions for the two free constants E_i and d,

$$E_i\, R = E\, R + \frac{d}{\epsilon_2 R^2}$$

$$\epsilon_1\, E_i = \epsilon_2 \left(E - \frac{2d}{\epsilon_2 R^3} \right).$$

From these relations it follows that

$$E_i = \frac{3}{\epsilon + 2}\, E \tag{2.59}$$

and

$$d = \epsilon_2\, \frac{\epsilon - 1}{\epsilon + 2}\, R^3\, E\,, \tag{2.60}$$

where the relative dielectric constant

$$\epsilon = \frac{\epsilon_1}{\epsilon_2}\,, \tag{2.61}$$

was introduced.

These solutions show that the field inside the dielectric sphere is reduced in comparison to that far away from the sphere by the *local field factor*

$$f = \frac{3}{\epsilon + 2}\,. \tag{2.62}$$

Eq. (2.60) shows that for its surrounding medium the dielectric sphere acts as a point dipole with the polarizability

$$\kappa = \frac{\epsilon - 1}{\epsilon + 2}\, \epsilon_2\, R^3\,. \tag{2.63}$$

As mentioned earlier, these results remain valid also in the case of electromagnetic waves whose wavelength is small in comparison to R, but the dielectric constants ϵ_1 and ϵ_2 have to be replaced by the frequency dependent complex dielectric functions $\epsilon_1(\omega)$ and $\epsilon_2(\omega)$, respectively. The dielectric function of the sphere depends on the radius through the confinement effects discussed in the different chapters of this book.

After this discussion of a single quantum dot let us consider the problem of several identical quantum dots which are homogeneously distributed inside the host material. The volume fraction of quantum-dot material is denoted by p. The macroscopically homogeneous distribution of the dots leads to a homogeneous macroscopic field E throughout the material. On the mesoscopic level, however, the field is different from the macroscopic one, depending on the spatial arrangement of the dots. Fortunately, there is a simple reasoning which allows to take into account the average effect of the mesoscopic inhomogeneity.

We have seen, that for the surrounding medium each dot behaves like a point-like polarizable dipole. Therefore, as long as the dot radii are smaller than the separation between the dots, we may identify the collective electrodynamics of this system with that of a system of point-like dipoles. This is just the standard model used in the classical atomic description of matter. Following the usual analysis, we consider one of the dipoles and circumscribe a sphere S_1 around it such that no other dipole exists inside that sphere. The field E_1 acting on this dipole is

$$E_1 = E + E_2 \ , \tag{2.64}$$

i.e., it consist of the macroscopic field E plus a supplementary field E_2 which results from the local deviation of the matter distribution from the average homogeneous situation.

For the field E_2 we now make again a macroscopic continuum approximation assuming that the matter is distributed continuously outside the sphere S_1, so that E_2 is given by the surface charge density on the sphere. The field E_2 in the center of the sphere has been computed e.g. by Sommerfeld [Sommerfeld (1949), Sec. 11.14.D] as

$$E_2 = \frac{4\pi}{3\epsilon_2} P \ , \tag{2.65}$$

where the integration has been evaluated over the surface of the sphere S_1. Within the polarizable dipole model one identifies the macroscopic polarization P with the induced dipole density

$$P = n \, d = n\kappa \, E_1 \ , \tag{2.66}$$

where n is the concentration and κ the polarizability of the dipoles. Inserting all this into Eq. (2.64) and eliminating the field E_1 in favor of the polarization **P** one finds

$$\mathbf{P} = \frac{(\epsilon_{eff} - 1)\epsilon_2}{4\pi} \mathbf{E} \tag{2.67}$$

with

$$\epsilon_{eff} = \frac{1 + \dfrac{8\pi}{3\epsilon_2} n\kappa}{1 - \dfrac{4\pi}{3\epsilon_2} n\kappa} \tag{2.68}$$

Therefore ϵ_{eff} is the effective macroscopic dielectric constant of the dipole plus dielectric cladding system relative to the dielectric constant ϵ_2 of the cladding. For $\epsilon_2 = 1$ (vacuum) one recognizes immediately the well-known Clausius-Mosotti relation. Its extension to propagating fields implies the introduction of a complex polarizability due to the possible absorption in the dipole. This variant of the formula is due to Lorentz-Lorenz [Born and Wolf (1968)].

For the case of a microsphere we now use Eq. (2.63) to express the dipole polarizability through the dielectric constant ϵ_1. This way we finally obtain the effective dielectric constant of the system as

$$\epsilon_{eff} = \frac{1 + 2p\,(\epsilon - 1)/(\epsilon + 2)}{1 - p\,(\epsilon - 1)/(\epsilon + 2)}$$

or

$$\epsilon_{eff} = 1 + \frac{3p\,(\epsilon - 1)}{(1-p)\,\epsilon + 2 + p}\;. \tag{2.69}$$

Here the concentration of the quantum dots (number of dots per unit volume) was replaced through their volume concentration (volume fraction)

$$p = \frac{4\pi R^3}{3} n\;.$$

This formula, for the case of propagating waves with complex dielectric function $\epsilon(\omega)$ instead of the real ϵ may be written also in the form

$$\frac{\epsilon_{eff}(\omega) - 1}{\epsilon_{eff}(\omega) + 2} = p\,\frac{\epsilon(\omega) - 1}{\epsilon(\omega) + 2} \tag{2.70}$$

and is known as the Maxwell-Garnett formula for composite media [Garnett (1904)].

If the system contains dots of different radii with the distribution $P(R/\bar{R})$ around their average radius $\bar{R}$ (see Sec. 2-3), the polarizability κ of the spheres in Eq. (2.66) should be replaced by their average polarizability $\bar{\kappa}$. Then Eq. (2.70) becomes

$$\frac{\epsilon_{eff}(\omega) - 1}{\epsilon_{eff}(\omega) + 2} = p \int d(R/\bar{R}) \, (R/\bar{R})^3 \, P(R/\bar{R}) \, \frac{\epsilon(\omega, R) - 1}{\epsilon(\omega, R) + 2} \, . \tag{2.71}$$

This equation is the basic relation to be used in the interpretation of the experiments on quantum dot systems. For many applications simplified versions of Eq. (2.71) are used. For small dot concentrations ($p \ll 1$) the first term of the expansion of $\epsilon_{eff}(\omega)$, Eq. (2.69), in powers of p gives

$$\epsilon_{eff}(\omega) \simeq 1 + 3p \, \frac{\epsilon(\omega) - 1}{\epsilon(\omega) + 2} = 1 + 3p - 9p \, \frac{1}{\epsilon(\omega) + 2} \, . \tag{2.72}$$

In the discussion of the absorption spectra one usually assumes that the real part of $\epsilon_{eff}(\omega)$ may be approximated by the background dielectric constant and only the frequency dependent structured contribution of the imaginary part is relevant. Moreover, it is assumed that the static real part of $\epsilon(\omega)$ is dominating since the bulk static dielectric constant of the dot material is about 10. Under these conditions also

$$\mathrm{Re} \; \epsilon_{eff}(\omega) \simeq 1$$

$$\tag{2.73}$$

$$\mathrm{Im} \; \epsilon_{eff}(\omega) \simeq \frac{9p}{(\epsilon + 2)^2} \, \mathrm{Im} \; \epsilon(\omega) \, .$$

The standard relation between the susceptibility and the complex dielectric function is

$$\epsilon_{eff}(\omega) = 1 + 4\pi \, \chi_{eff}(\omega)$$

and the absorption coefficient α is then proportional to $\mathrm{Im} \; \chi(\omega)$. Taking into account the radius distribution we have

$$\alpha(\omega) \propto p \int d(R/\bar{R}) \, (R/\bar{R})^3 \, P(R/\bar{R}) \; \text{Im} \, \chi_{eff}(\omega, R) \; . \qquad (2.74)$$

One additional complication in the interpretation of this formula can arise in the case of very strong fields, where the dielectric function $\epsilon(\omega)$ of a dot might depend also on the intensity of the field inside the dot. This complication is connected with the fact that the field E_i inside the dot does not coincide with the field E outside the dot. Even if the dependence of $\epsilon(\omega)$ on the field intensity in the dot $I_i \propto |E_i|^2$ is (theoretically) known, then it is still a problem to find its dependence on the intensity I. For illustration let us assume that $\epsilon(\omega)$ has the intensity dependence given by the function $g(\omega, I_i)$, so that

$$\epsilon(\omega) = g(\omega, I_i) \; , \qquad (2.75)$$

where I_i and I are related through the absolute square of Eq. (2.59). Then we have to solve the implicit equation [Chemla and Miller (1986), Leung (1986), Schmitt-Rink et al. (1987)]

$$\epsilon(\omega, I) = g\left(\omega, \; \frac{9I}{\left|\epsilon(\omega, I) + 2\right|^2}\right) \qquad (2.76)$$

to get the dependence of the dielectric function on the intensity I. This relation is not necessarily single valued, and optical bistability based on this effects has been proposed, but so far could not be observed experimentally.

2-5. Surface Polarization and Particle Interactions

The influence of Coulomb effects on the electron–hole excitations is one of the central aspects of the discussion in this book. For this purpose we need the correct analysis of the electrostatic properties of the quantum dot in its surrounding material. Already in the previous section we discussed the case of a homogeneous field. This section is now devoted to the classical electrostatics of the inhomogeneous field created by the presence of point charges.

Within the frame of classical electrostatics the Coulomb interaction energy of two point-charges q_1, q_2 at r_1, r_2 in an infinite dielectric medium with background dielectric constant ϵ_1 is

$$W_{12} = \frac{q_1 q_2}{\epsilon_1 |r_1 - r_2|} .$$
(2.77)

The background dielectric constant ϵ_1 describes the screening of the bare Coulomb interaction through the macroscopic polarization of the medium, including both electronic and ionic contributions.

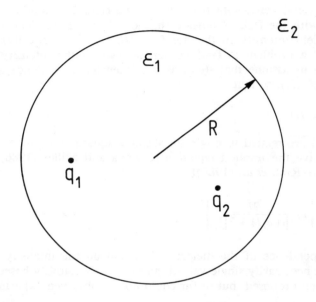

Fig. 2.6 Point charges in a dielectric sphere.

The external electrostatic field induces surface polarization charges at the interface between two media with different dielectric constants. In the presence of such interfaces, one also has to consider the effect of the surface polarization on the effective Coulomb interaction between all other charges. The underlying idea is to separate between charges introduced arbitrarily into the system (external charges) and charges which are induced as consequence of the dielectric medium properties. In our approach, these induced charges are always treated in the framework of macroscopic electrostatics. The goal is to describe the energy of the system as function of the positions of the external charges. This problem has an analytical solution only for certain simple geometries. The example of two external charges inside a sphere of radius R with dielectric constant ϵ_1, which is embedded in an infinite medium with dielectric constant ϵ_2 has been solved by Brus (1984).

We give here a derivation that includes the cases where *i*) both external charges are inside the sphere, *ii*) both charges are outside, and *iii*) one charge is inside and one charge is outside, respectively. We first derive the electrostatic potential V of a single external unit charge ($q = 1$) inside (outside) the sphere. Let us denote by **s** the position of the charge inside (outside) the sphere in a reference system with the center of the sphere as the origin. We want to find the electrostatic potential at an arbitrary point **r** inside or outside the sphere, as sketched in Fig. 2.6. For this purpose we solve Poisson's equation with the proper boundary conditions. Inside the sphere we have

$$\epsilon_1 \nabla^2 V(\mathbf{r}) = -4\pi \rho(\mathbf{r}) \qquad (r < R) , \qquad (2.78)$$

and outside the sphere

$$\epsilon_2 \nabla^2 V(\mathbf{r}) = -4\pi \rho(\mathbf{r}) \qquad (r > R) . \qquad (2.79)$$

The charge density

$$\rho(\mathbf{r}) = \delta(\mathbf{r} - \mathbf{s}) \qquad (2.80)$$

vanishes outside (inside) the sphere, for $s < R$ ($s > R$). The boundary conditions imply the continuity of the tangential component of the electric field,

$$\mathbf{r} \times \nabla V \bigg|_{r = R - 0} = \mathbf{r} \times \nabla V \bigg|_{r = R + 0} , \qquad (2.81)$$

and of the normal component of the induction field

$$\epsilon_1 \mathbf{r} \cdot \nabla V \bigg|_{r = R - 0} = \epsilon_2 \mathbf{r} \cdot \nabla V \bigg|_{r = R + 0} . \qquad (2.82)$$

We look for the solution of this problem as the superposition of a particular solution of the inhomogeneous problem and the general solution of the homogeneous problem. For the inhomogeneous solution we choose one of the bulk solutions ($R = \infty$ respectively $R = 0$)

$$V_c(\mathbf{r}) = \frac{1}{|\mathbf{r} - \mathbf{s}|} \begin{cases} \dfrac{1}{\epsilon_1} & \text{for } s < R \\[2ex] \dfrac{1}{\epsilon_2} & \text{for } s > R \end{cases} \qquad (2.83)$$

Consequently, the full solution is

$$V(\mathbf{r}) = V_c(\mathbf{r}) + \delta V(\mathbf{r}) \ , \qquad (2.84)$$

where δV has to satisfy the homogeneous Laplace equation.

The symmetry of the problem implies that, besides its s-dependence, the potential V depends only on r and the angle θ between the vectors $\mathbf{r}$ and $\mathbf{s}$. Hence, it is advantageous to use a decomposition in terms of Legendre polynomials,

$$V(\mathbf{r}) = \sum_{\ell = 0}^{\infty} V_\ell(r) \, P_\ell(\cos\theta) \ . \qquad (2.85)$$

For the Coulomb potential in bulk semiconductors we use the partial-wave decomposition,

$$\frac{1}{|\mathbf{r} - \mathbf{s}|} = \frac{1}{\epsilon_1} \sum_{\ell = 0}^{\infty} \left[\frac{1}{r} \left(\frac{r}{s} \right)^{\ell+1} \theta(s-r) + \frac{1}{s} \left(\frac{s}{r} \right)^{\ell+1} \theta(r-s) \right] P_\ell(\cos\theta) \ .$$

$$(2.86)$$

The ℓ-th partial wave δV_ℓ of the homogeneous solution δV satisfies the homogeneous Laplace equation

$$\frac{1}{r} \frac{\partial^2}{\partial r^2} [r \, \delta V_\ell(r)] - \frac{\ell(\ell+1)}{r^2} \delta V_\ell(r) = 0 \ . \qquad (2.87)$$

This equation has two independent solutions,

$$r^\ell \quad \text{and} \quad r^{-\ell-1} \ .$$

The behavior at $r = 0$ and $r = \infty$ imposes the choice of the solution inside and outside the sphere

$$\delta V_\ell(r) = \begin{cases} A_\ell \, r^\ell & \text{for } r < R \\ B_\ell \, r^{-\ell-1} & \text{for } r > R \end{cases} \tag{2.88}$$

The two constants A_ℓ and B_ℓ are determined from the boundary conditions for the partial waves,

$$\epsilon_1 \left.\frac{\partial V_\ell}{\partial r}\right|_{r = R - 0} = \epsilon_2 \left.\frac{\partial V_\ell}{\partial r}\right|_{r = R + 0} \tag{2.89}$$

and

$$V_\ell\Big|_{r = R - 0} = V_\ell\Big|_{r = R + 0}. \tag{2.90}$$

Note, that actually the continuity of the $\ell = 0$ partial wave is not necessary but convenient. The potential is determined only up to a constant.

In summary, the complete solution is

$$V(\mathbf{r};\, \mathbf{s}) = \frac{1}{\epsilon_1}\left\{ \frac{1}{|\mathbf{r} - \mathbf{s}|} + \frac{\epsilon - 1}{R} \sum_{\ell=0}^{\infty} \left(\frac{s}{R}\right)^\ell \left(\frac{r}{R}\right)^{\alpha_\ell} P_\ell\left(\frac{\mathbf{r}\cdot\mathbf{s}}{rs}\right) \frac{\ell+1}{1+\ell(\epsilon+1)} \right\} \tag{2.91}$$

for the charge inside the sphere ($s < R$), and

$$V(\mathbf{r};\, \mathbf{s}) = \frac{1}{\epsilon_2}\left\{ \frac{1}{|\mathbf{r} - \mathbf{s}|} - \frac{\epsilon - 1}{R} \sum_{\ell=0}^{\infty} \left(\frac{s}{R}\right)^{-\ell-1} \left(\frac{r}{R}\right)^{\alpha_\ell} P_\ell\left(\frac{\mathbf{r}\cdot\mathbf{s}}{rs}\right) \frac{\ell}{1+\ell(\epsilon+1)} \right\} \tag{2.92}$$

for the charge outside the sphere ($s > R$), respectively. Here, we defined

$$\alpha_\ell = \begin{cases} \ell & \text{for } r < R \\ -\ell-1 & \text{for } r > R \end{cases}, \tag{2.93}$$

and ϵ is given by Eq. (2.61).

We now turn our attention to the electrostatic energy of a system of point charges distributed inside and outside the dielectric sphere. According to classical electrostatics the energy is given by

$$W = \frac{1}{8\pi} \int d^3r \; \mathbf{E}(\mathbf{r}) \cdot \mathbf{D}(\mathbf{r}) \; , \tag{2.94}$$

where $\mathbf{E}$ and $\mathbf{D}$ are the electric field and the induction field, respectively. The integration extends over the whole space. Introducing the electrostatic potential V

$$\mathbf{E}(\mathbf{r}) = - \boldsymbol{\nabla} \, V(\mathbf{r}) \; , \tag{2.95}$$

and using Poisson's equation

$$\boldsymbol{\nabla} \cdot \mathbf{D}(\mathbf{r}) = - 4\pi \, \rho(\mathbf{r}) \; , \tag{2.96}$$

we may perform a partial integration to eliminate the fields $\mathbf{E}$ and $\mathbf{D}$ in favor of V and ρ. However, we have to take into account that the $\mathbf{E}$ field is discontinuous at the interface between the two media. If we choose a continuous potential which vanishes at infinity, we have

$$W = \frac{1}{2} \int d^3r \; V(\mathbf{r}) \, \rho(\mathbf{r}) \; . \tag{2.97}$$

The charge density which describes N point-charges q_i, $i = 1, 2, \dots N$, is

$$\rho(\mathbf{r}) = \sum_{i=1}^{N} q_i \, \delta(\mathbf{r} - \mathbf{s}_i) \; . \tag{2.98}$$

Using the superposition principle we find the total potential of the N charges as the sum of the potentials created by each charge as

$$V(\mathbf{r}) = \sum_{i=1}^{N} q_i \, V(\mathbf{r}; \mathbf{s}_i) \; , \tag{2.99}$$

where $V(\mathbf{r}; \mathbf{s}_i)$ is the potential calculated before.

Due to the infinite self-energy of a classical point charge the resulting energy W is divergent. However, the variation of the self-energy due to the difference in the dielectric constants ($\epsilon \neq 1$) is well defined. We therefore define the effective Coulomb interaction energy W_{eff} by subtracting the divergent contributions

$$\lim_{r \to 0} \frac{q_i^2}{\epsilon_1 r} \quad \text{or} \quad \lim_{r \to 0} \frac{q_i^2}{\epsilon_2 r}$$

for a point charge q_i inside or outside the sphere, respectively. One should note that these divergent contributions are also eliminated in the same way from the electrostatic theory of bulk materials.

For our quantum-dot problem we get

$$W_{eff} = \frac{1}{2} \sum_{\substack{i=1 \\ i \neq j}}^{N} q_i \, q_j \, V(\mathbf{s}_i; \mathbf{s}_j) + \frac{1}{2} \sum_{i=1}^{N} q_i^2 \, \delta V(\mathbf{s}_i; \mathbf{s}_i) \, . \tag{2.100}$$

The potentials $V(\mathbf{s}_i; \mathbf{s}_j)$ and $\delta V(\mathbf{s}_i; \mathbf{s}_i)$ are defined through Eqs. (2.83) - (2.84) and (2.91) - (2.92), respectively.

A typical feature of the effective Coulomb energy in the presence of the dielectric surface polarization is the survival of the finite, coordinate dependent self-energy terms $\delta V(\mathbf{s}_i; \mathbf{s}_i)$, which are unbounded close to the surface of the sphere. Indeed, the partial-wave summation in Eq. (2.91) for $\mathbf{r} = \mathbf{s}$ and $s \to R$ (from $s < R$) diverges as the geometrical series

$$\frac{\epsilon - 1}{2(\epsilon + 1)\epsilon_1} \frac{1}{1 - r/R} \, . \tag{2.101}$$

Similarly, in Eq. (2.92) we get the asymptotic behavior for $s \to R$ (from $s > R$)

$$\frac{\epsilon - 1}{2(\epsilon + 1)\epsilon_2} \frac{1}{1 - r/R} \, . \tag{2.102}$$

Note, that due to the difference of the coefficients in Eqs. (2.101) and (2.102), the singularity at the surface is not integrable in the sense of a principle value integral. The same kind of behavior is obtained also in the case of a plane parallel slab configuration [Tran Thoai et al. (1990)]. These unphysical results are consequences of the assignment of the surface polarization to an idealized mathematical dipole surface. We will come back to this problem in Sec. 4-4 where we discuss a regularized surface potential and its influence on the electron-hole-pair states in a quantum dot.

In most of the discussion in this book we assume the idealized case where all the particles are inside the sphere. In this configuration, if the system is neutral,

$$\sum_{i=1}^{N} q_i = 0 \ , \tag{2.103}$$

we may write

$$W_{eff} = \sum_{\substack{i,j=1 \\ i>j}}^{N} \left[\frac{q_i q_j}{\epsilon_1} \frac{\epsilon-1}{R} \left\{ \sum_{\ell=1}^{\infty} \left(\frac{r_i r_j}{R^2} \right)^{\ell} P_\ell \left(\frac{\mathbf{r}_i \cdot \mathbf{r}_j}{r_i r_j} \right) \frac{1}{1+\epsilon\ell/(\ell+1)} \right. \right.$$

$$\left. \left. + \frac{1}{2} \sum_{k=1}^{\infty} [(r_i/R)^{2k} + (r_j/R)^{2k}] \frac{1}{1+\epsilon k/(k+1)} \right\} + \frac{q_i q_j}{\epsilon_1 |\mathbf{s}_i - \mathbf{s}_j|} \right] \ . \tag{2.104}$$

The disappearance of the $\ell = 0$ terms here is a consequence of charge neutrality.

Chapter 3
QUANTUM CONFINEMENT REGIMES

In their pioneering investigations of quantum confinement in semiconductor microcrystallites, Efros and Efros (1982) introduced three regimes of quantum confinement, depending on the ratio of the crystallite radius R to the Bohr radius of the electrons, holes, and electron-hole pairs, respectively. For very small quantum dots one speaks of the *strong confinement regime*, where the individual motions of the electron and the hole are quantized.

For somewhat larger dots one can introduce an *intermediate confinement regime* if the effective mass of the holes is much bigger than that of the electrons. In this case the radius of the microsphere has to be small relative to the electron Bohr radius and large relative to the hole Bohr radius, respectively. Under these conditons the influence of the quantum confinement on electron and hole is substantially different.

The *weak confinement regime* is appropriate for relatively large quantum dots. In the theoretical analysis of the optical and electronic properties of such systems one often introduces relative and center-of-mass coordinates of the electron-hole pairs, exactly as in the case of a bulk semiconductor material. Furthermore, it is typical to assume that the quantum confinement effects do not interfere with the relative motion of the pair, leading only to a quantized center-of-mass motion.

We discuss the three quantum confinement regimes in the different sections of this chapter. In each section we perform a scaling analysis of the appropriate electron-hole-pair Hamiltonian and introduce approximation schemes which are best suited for the respective situations.

3-1. Strong Confinement

In Chap. 2 we showed that a large confinement induced energy shift occurs for quantum dots with a small radius,

$$R \ll a_B \ . \tag{3.1}$$

For these dots it has been speculated in the literature that it might be a good approximation to completely ignore the Coulomb interaction since the movement of the carriers is strongly quantized in all spatial directions. This is the basic assumption behind the *strong confinement approximation*. Even though we discuss this ideal strong confinement limit in the current section, we want to mention already at this point that more elaborate numerical calculations (Sec. 4-2) have shown that it is actually never justified to ignore the Coulomb interaction. Nevertheless, the strong confinement limit is still a relevant problem, since it can be solved analytically for perfect confinement conditions. Furthermore, the solutions are very helpful as a complete expansion set for numerical calculations which include the Coulomb interaction potential and possibly also other complications of real microcrystallites.

To analyze the strong confinement limit and corrections to it, we consider a system of electrons and holes (without spin) in the parabolic, isotropic effective mass approximation confined in a sphere of radius R. Since electrons and holes are always created or annihilated pairwise in optical experiments, we may restrict our discussion to systems with an identical number N of electrons and holes. The Hamiltonian describing this system is

$$\mathcal{H}^{(N)} = T^{(N)}(\mathbf{r}) + W^{(N)}(\mathbf{r}_e, \mathbf{r}_h) \tag{3.2}$$

with the kinetic energy

$$T^{(N)}(\mathbf{r}) = - \sum_{i=1}^{N} \left[\frac{\hbar^2}{2m_e} \nabla_{e_i}^2 + \frac{\hbar^2}{2m_h} \nabla_{h_i}^2 \right] \tag{3.3}$$

and the potential energy

$$W^{(N)}(\mathbf{r}_e, \mathbf{r}_h) = \frac{1}{2} \sum_{\substack{i, j = 1 \\ i \neq j}}^{N} [W(\mathbf{r}_{e_i}, \mathbf{r}_{e_j}) + W(\mathbf{r}_{h_i}, \mathbf{r}_{h_j})]$$

$$- \sum_{i, j = 1}^{N} W(\mathbf{r}_{e_i}, \mathbf{r}_{h_j}) . \qquad (3.4)$$

Here $W(\mathbf{r}_i, \mathbf{r}_j)$ is the effective Coulomb interaction energy of two particles of charge e in the dielectric sphere. (Note that we drop the index *eff* to lighten the notation.) The eigenstates of the confined electron-hole system have to satisfy the boundary conditions

$$\Psi^{(N)}(\mathbf{r}_{e_i}, \mathbf{r}_{h_i}) \bigg|_{r_i = R} = 0 \quad \text{for all} \quad i = 1, 2, ..., N . \qquad (3.5)$$

It is useful to move the radius dependence from this boundary condition completely into the Hamiltonian by scaling the space coordinates in units of the dot radius R. Furthermore, we use again

$$\frac{\hbar^2}{2m_r R^2} = E_R \left(\frac{a_B}{R} \right)^2 \qquad (3.6)$$

as unit of the energy. We introduce the dimensionless parameter

$$\lambda = \frac{R}{a_B} , \qquad (3.7)$$

so that after the rescaling of the coordinates,

$$\mathbf{r} = R \mathbf{x} , \qquad (3.8)$$

the kinetic energy scales as

$$T(\mathbf{r}) = E_R \lambda^{-2} T(\mathbf{x}) , \qquad (3.9)$$

where $T(\mathbf{x})$ depends only on the ratio of the electron and hole masses. The effective Coulomb interaction energy scales as

$$W(\mathbf{r}_e, \mathbf{r}_h) = E_R \, \lambda^{-1} \, W(\mathbf{x}_e, \mathbf{x}_h) \tag{3.10}$$

where the function

$$W(\mathbf{x}, \mathbf{y}) = \frac{2}{|\mathbf{x} - \mathbf{y}|} + 2\,(\epsilon - 1) \tag{3.11}$$

$$\times \left[\sum_{\ell=1}^{\infty} (xy)^{\ell} \, P_{\ell}\!\left[\frac{\mathbf{x}\cdot\mathbf{y}}{xy}\right] \frac{1}{1 + \epsilon\ell/(\ell+1)} + \frac{1}{2} \sum_{k=1}^{\infty} (x^{2k} + y^{2k}) \, \frac{1}{1 + \epsilon k/(k+1)} \right]$$

is a universal function, that does not depend on any other parameter except ϵ, the ratio of the two dielectric constants.

Hence, we scale the total Hamiltonian as

$$\mathcal{H}^{(N)}(\mathbf{r}) = E_R \, \lambda^{-2} \, \mathcal{H}^{(N)}(\mathbf{x}) , \tag{3.12}$$

where

$$\mathcal{H}^{(N)}(\mathbf{x}) = - \sum_{i=1}^{N} \left(\frac{m_r}{2m_e} \nabla_{e_i}^2 + \frac{m_r}{2m_{h_i}} \nabla_{h_i}^2 \right)$$

$$+ \lambda \left\{ \frac{1}{2} \sum_{\substack{i,j=1 \\ i \neq j}}^{N} [\, W(\mathbf{x}_{e_i}, \mathbf{x}_{e_j}) + W(\mathbf{x}_{h_i}, \mathbf{x}_{h_j})] - \sum_{i,j=1}^{N} W(\mathbf{x}_{e_i}, \mathbf{x}_{h_j}) \right\}. \tag{3.13}$$

The boundary conditions for the wavefunctions in the scaled coordinates now are

$$\Psi^{(N)}(\mathbf{x}_{e_i}, \mathbf{x}_{h_i}) \Big|_{x_i = 1} = 0 \quad \text{for all } i = 1, 2, ..., N . \tag{3.14}$$

In Eq. (3.13) we see directly that for $\lambda \ll 1$ the Coulomb interaction acts as a perturbation to the kinetic energy contribution. In other words, the kinetic energy is the dominant energy contribution for small radii $R \ll a_B$. The fact that the true kinetic energy diverges for $R \to 0$ is included in the diverging rescaling of the Hamiltonian.

One of the successful methods to deal with problems having a discrete perturbed and unperturbed spectrum, as in our case, is stationary (analytic) perturbation theory. Standard nondegenerate perturbation theory yields the perturbed ground-state energy E up to second order in λ as

$$E = E_0 + \lambda \langle 0|\mathscr{H}'|0\rangle + \lambda^2 \sum_i \frac{|\langle 0|\mathscr{H}'|i\rangle|^2}{E_0 - E_i} . \tag{3.15}$$

Here $|0\rangle$ and E_0 are the unperturbed state vector and energy, $\mathscr{H}'$ is the perturbation Hamiltonian, while $|i\rangle$ and E_i are the unperturbed state vectors and energies of all the other states, respectively. More precisely, in our case $|0\rangle$ is the ground state of the kinetic energy operator for the N-pair electron-hole system in the unit sphere, and the perturbation is the sum of all effective (rescaled) Coulomb interactions.

In what follows we consider only the one- (exciton) and two-electron-hole-pair (biexciton) problems. In the absence of spin both ground states are nondegenerate. Introduction of spin will not affect our results, except for a supplementary spin degeneracy. This spin degeneracy, however, does not affect the applicability of the nondegenerate perturbation theory, since at the current level of approximation the spin degrees of freedom are decoupled from the orbital ones. In the case of two electrons and holes the relevant state might be interpreted as that where the two electrons (holes) have opposite spins. The biexciton ground state in bulk semiconductors has this configuration. These features allow us to ignore complications due to identical particles.

The eigenfunctions of the unperturbed problem are simply the products of the one-particle kinetic energy eigenfunctions, Eqs. (2.15) – (2.17), which in the rescaled notation are

$$\psi_{n\ell m}(\mathbf{x}) = \phi_{n\ell m}(x) \, Y_{\ell m}(\theta,\phi)$$

with

$$\phi_{n\ell m}(x) = \sqrt{2} \, \frac{j_\ell(\kappa_{n,\ell}x)}{j_{\ell+1}(\kappa_{n,\ell})} . \tag{3.16}$$

These eigenfunctions do not depend on the masses and, hence, are the same for electrons and holes. As discussed in Chap. 2, the energy eigenvalues are determined from $j_\ell(\kappa_{n,\ell}) = 0$ and are found to be inversely proportional to the mass of the particle,

$$E^\alpha_{n,\ell} = \frac{m_r}{m_\alpha} \kappa^2_{n,\ell} \quad \text{for } \alpha = e, h. \tag{3.17}$$

For our discussion of not too large quantum dots in this and later chapters, it will be sufficient to deal with the one- and two-electron-hole-pair problem. For this purpose, we introduce the notation

$$\mathscr{H}_x = \mathscr{H}^{(1)} \tag{3.18}$$

and

$$\mathscr{H}_{xx} = \mathscr{H}^{(2)} , \tag{3.19}$$

for the one-pair (exciton) and two-pair (biexciton) Hamiltonian, respectively. The detailed expressions for $\mathscr{H}^{(1)}$ and $\mathscr{H}^{(2)}$ follow directly from Eqs. (3.2) - (3.4).

The relevant matrix element for the one electron-hole-pair problem is

$$\langle 0,0 | \mathscr{H}'_x | \alpha, \beta \rangle = - \int d^3x_e \, d^3x_h \, \psi_0(\mathbf{x}_e) \, \psi_0(\mathbf{x}_h) \, W(\mathbf{x}_e, \mathbf{x}_h) \, \psi_\alpha(\mathbf{x}_e) \, \psi_\beta(\mathbf{x}_h)$$

$$= - V_{\alpha\beta} = - V_{\beta\alpha} , \tag{3.20}$$

where in a short-hand notation α (or β) stands for a given set of one-particle-state quantum numbers (n,ℓ,m) in the sphere. Zero, (0 or $_0$ as index) stands for the set $(1,0,0)$ describing the one-particle ground state.

In the two electron-hole-pair problem the relevant matrix element can be expressed as

$$\langle 00,00 | \mathscr{H}'_{xx} | \alpha\alpha', \beta\beta' \rangle = V_{\alpha\alpha'} \, \delta_{\beta 0} \, \delta_{\beta' 0} + V_{\beta\beta'} \, \delta_{\alpha 0} \, \delta_{\alpha' 0}$$

$$- V_{\alpha\beta} \, \delta_{\alpha' 0} \, \delta_{\beta' 0} - V_{\alpha'\beta'} \, \delta_{\alpha 0} \, \delta_{\beta 0} - V_{\alpha\beta'} \, \delta_{\alpha' 0} \, \delta_{\beta 0} - V_{\alpha'\beta} \, \delta_{\alpha 0} \, \delta_{\beta' 0} . \tag{3.21}$$

It is convenient to count the one-particle energies from the lowest one, $e_0^{e,h}$, and to denote

$$\epsilon_\alpha^{e,h} = e_\alpha^{e,h} - e_0^{e,h} \ . \tag{3.22}$$

With this notation, the first two terms of our perturbation theory for the energy of the electron-hole-pair (exciton) ground state in the energy units of Eq. (3.6) are,

$$E_x = e_0^e + e_0^h - \lambda V_{00} - \lambda^2 \sum_{\alpha\beta}{}' \frac{|\langle 0,0|\mathcal{H}'_x|\alpha,\beta\rangle|^2}{\epsilon_\alpha^e + \epsilon_\beta^h}$$

$$= e_0^e + e_0^h - \lambda V_{00} - \lambda^2 \sum_{\alpha\beta}{}' \frac{|V_{\alpha\beta}|^2}{\epsilon_\alpha^e + \epsilon_\beta^h} \ . \tag{3.23}$$

Here, the prime on the summation indicates that the $\alpha = \beta = 0$ term has to be omitted. Correspondingly, the two-electron-hole-pair (biexciton) ground state energy is

$$E_{xx} = 2\,(e_0^e + e_0^h - \lambda V_{00}) - \lambda^2 \sum_{\alpha\alpha'\beta\beta'} \frac{|\langle 00,00|\mathcal{H}'_{xx}|\alpha\alpha',\beta\beta'\rangle|^2}{\epsilon_\alpha^e + \epsilon_{\alpha'}^e + \epsilon_\beta^h + \epsilon_{\beta'}^h}$$

$$= 2\,(e_0^e + e_0^h - \lambda V_{00}) - \lambda^2 \sum_{\alpha\beta}{}' |V_{\alpha\beta}|^2 \left(\frac{4}{\epsilon_\alpha^e + \epsilon_\beta^h} + \frac{1}{\epsilon_\alpha^e + \epsilon_\beta^e} \right.$$

$$\left. + \frac{1}{\epsilon_\alpha^h + \epsilon_\beta^h} \right) + 4\lambda^2 \sum_\alpha{}' |V_{\alpha 0}|^2 \left(\frac{1}{\epsilon_\alpha^e} + \frac{1}{\epsilon_\alpha^h} \right) \ . \tag{3.24}$$

The prime on the last summation indicates that the $\alpha = 0$ term has to be omitted.

We now have to remember that the actual energies are obtained from Eqs. (3.23) and (3.24) through multiplication with $E_R \lambda^{-2}$. Consequently, the kinetic energy terms scale like λ^{-2}, the linear Coulomb terms scale like λ^{-1}, and the quadratic Coulomb terms have no λ prefactor. Hence, for small λ the kinetic energy dominates, but the other terms may also be important, for example in the typical situation when the energies of two different states are degenerate in the absence of the Coulomb interaction.

Let us consider here the *molecular binding energy* (*biexciton binding energy*), defined as

$$E_{mol} = 2 E_x - E_{xx} .$$ (3.25)

In bulk semiconductors this energy combination corresponds to the binding energy of two excitons to form a molecule (biexciton), since the two-exciton state is also an eigenstate of the same Hamiltonian. It is possible to arbitrarily increase the distance between two excitons, which leads to an arbitrary reduction of their interaction energy. In quantum dots, however, one cannot truly dissociate a biexciton since the electrons and holes are confined inside the dot. Consequently, the energy $2 E_x$ is not a true eigenenergy of the system. Nevertheless, the energy difference (3.25) is a meaningful quantity to study, in particular since it may be obtained experimentally, e.g., by comparing measurements using one- and two-photon spectroscopy. According to Eqs. (3.23) - (3.24) E_{mol} vanishes in zeroth- and first-order perturbation theory and has a finite, λ independent, strictly positive value in second-order perturbation theory,

$$E_{mol}^{(2)} = E_R \sum_{\alpha, \beta > 0} |V_{\alpha\beta}|^2 \left(\frac{2}{\epsilon_\alpha^e + \epsilon_\beta^h} + \frac{1}{\epsilon_\alpha^e + \epsilon_\beta^e} + \frac{1}{\epsilon_\alpha^h + \epsilon_\beta^h} \right) .$$ (3.26)

Actually, due to the λ-dependence of the perturbation theory result, Eq. (3.26) gives the $\lambda \to 0$ ($R \ll a_B$) limit of the molecular binding energy as

$$\lim_{\lambda \to 0} E_{mol} = E_{mol}^{(2)} .$$ (3.27)

To obtain an estimate for this energy, it is sufficient to note that the sum in Eq. (3.26) contains only positive terms, and therefore the first term gives already a lower bound. The lowest unperturbed excited state which contributes to the summation in Eq. (3.26) is the degenerate state corresponding to $\alpha = \beta = 1$, i.e. $\ell = 1$, $m = 0, \pm 1$, $n = 1$. This state is associated with the one-particle energy increment

$$\epsilon_1^{e,h} = \frac{m_r}{m_{e,h}} (\kappa_{1,1}^2 - \kappa_{0,1}^2) \simeq 10.32 \frac{m_r}{m_{e,h}} .$$ (3.28)

Therefore,

$$\lim_{\lambda \to 0} E_{mol} > E_R \left[2 + \frac{m_e + m_h}{m_r} \right] C(\epsilon) , \qquad (3.29)$$

where the constant $C(\epsilon)$ depends only on the ratio ϵ of the dielectric constants,

$$C(\epsilon) = \frac{1}{10.32} \sum_{m, m' = 0, \pm 1} |V_{1m1,1m'1}|^2 . \qquad (3.30)$$

A simple numerical calculation of the corresponding matrix elements for different values of ϵ gives

$$C(1) = 0.052 ; \quad C(4) = 0.076 ; \quad C(10) = 0.104 . \qquad (3.31)$$

Hence, for an electron-hole mass ratio of $m_e/m_h = 0.1$ and a dielectric constant ratio $\epsilon = 10$ already this modest lower bound gives a quite large asymptotic molecular binding energy exceeding the bulk exciton Rydberg energy.

It is important here to note the role of the dielectric polarization effects which increased this lower bound for the binding energy by a factor of 2 for $\epsilon = 10$ as compared to $\epsilon = 1$. The situation with $\epsilon = 10$ may be encountered for semiconductor quantum dots embedded in glass or in a liquid. Although these results have been obtained for a highly idealized limit, $R/a_B \to 0$ [Bányai (1989)] they may serve as a guide for approximate numerical calculations for finite but small radii. Furthermore, it gives a good estimate [Hu et al. (1990a)] already for $R/a_B < 1$, as we shall see in Chap. 4.

Besides the energies the Coulomb interaction affects also the wavefunctions. These modifications may also be treated within the frame of non-degenerate or degenerate perturbation theory. The specific feature of these perturbative corrections is that through the mixing of the unperturbed states, the simple selection rules of Chap. 2 are no longer appropriate. Actually, one has here a reduction of the symmetry of the Hamiltonian. In the absence of Coulomb interactions the invariance against independent rotations of each particle around the center of the spherical dot is assured, whereas in the presence of Coulomb interactions there is only an invariance against simultaneous rotations of all particles around the dot center.

Another consequence of the Coulomb interaction is that the electron-hole symmetry in the wavefunctions disappears so that the wavefunctions depend directly on the electron-hole mass ratio. Consequently, the charge distribution of the electrons is different from that of the holes, leading to a non-vanishing local charge density. The appearance of this local charge density may have far reaching consequences in polar crystals. In the following chapters we come back to problems related to the influence of Coulomb interactions on the wavefunctions, using non-perturbative approaches.

The perturbative method of expansion in powers of a small parameter, characterizing the dimension of the quantum dot relative to the exciton Bohr radius a_B, can be applied not only to microspheres but also to other quantum dot geometries. However, for other geometries the dielectric corrections have not been computed and have therefore never been taken into account. Fortunately, the case of strongly different dielectric constants ($\epsilon \gg 1$) occurs mainly for microcrystallites embedded in glass or in liquids, where dominantly spherical geometries occur. Even more, in this case, due to the large difference in the energy gaps of the surrounding material and the microcrystallites, the height of the potential barrier can sometimes be taken as infinite, as we did in this chapter. The vanishing of the wavefunctions at the boundary, and the fact that $\epsilon > 1$, also eliminates potential complications related to the pathology of the Coulomb self-energy mentioned in Chap. 2.

In the case of quantum dots fabricated from narrow quantum wells using nanolithography, the simplest geometry to be considered is a layer of very small thickness w having the shape of a square box of lateral dimension A. Here, we restrict ourselves to the idealization of zero thickness ($w = 0$), infinitely high potential barriers, and vanishing dielectric boundary effects. The quantum mechanical problem of N electron-hole pairs in this box is described by the Hamiltonian

$$
\mathcal{H}^{(N)} = - \sum_{i=1}^{N} \left(\frac{\hbar^2}{2m_e} \nabla^2_{e_i} + \frac{\hbar^2}{2m_h} \nabla^2_{h_i} \right)
$$
$$
+ \frac{e^2}{2\epsilon_1} \sum_{\substack{i,j=1 \\ i \neq j}}^{N} \left[\frac{1}{|\mathbf{r}_{e_i} - \mathbf{r}_{e_j}|} + \frac{1}{|\mathbf{r}_{h_i} - \mathbf{r}_{h_j}|} \right] - \frac{e^2}{\epsilon_1} \sum_{i,j=1}^{N} \frac{1}{|\mathbf{r}_{e_i} - \mathbf{r}_{h_j}|} .
$$

$$(3.32)$$

The coordinates, and implicitly the Laplace operators, are two-dimensional $\mathbf{r} = (x, y)$, and restricted to the box $|x|, |y| < A/2$. The many-particle wavefunction has to satisfy the boundary conditions

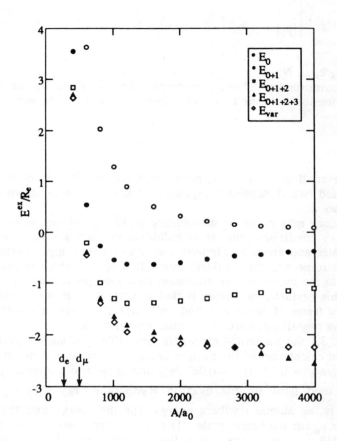

Fig. 3.1 The ground-state energy of the exciton in a two-dimensional box within different orders of perturbation theory [Bryant (1990)] as a function of the lateral size A in nm. The results of a variational calculation [Bryant (1988)] are also shown. (m_e/m_h = 0.75)

$$\Psi^{(N)}(x_i, y_i) \bigg|_{|x_i| = A/2} = 0$$

$$\Psi^{(N)}(x_i, y_i) \Bigg|_{|y_i| = A/2} = 0, \qquad\qquad (3.33)$$

for all $i = 1, 2, ..., N$.

The appropriate smallness parameter λ here is given by the ratio of the box dimension A to the two-dimensional exciton Bohr diameter d_μ,

$$\lambda = \frac{A}{d_\mu}.$$

Note, however, that for λ much smaller than one the validity of the model with idealized two-dimensional geometry itself might break down since it assumes $A \gg w$.

Again, one may move the dependence on the lateral dimension A from the boundary conditions into the Hamiltonian through a suitable rescaling of coordinates and energies. Indeed, one may naively apply perturbation theory with respect to the Coulomb interaction terms, which is completely equivalent to the more rigorous treatment described previously. The peculiarity of this perturbation theory is that it starts with terms of order λ^{-2}, followed by terms of order λ^{-1}, and only thereafter with a whole Taylor series in non negative powers of the small parameter λ.

In Fig. 3.1 the numerical results [Bryant (1990)] of such a third-order perturbation calculation of the exciton energy are shown for the mass parameters $m_e/m_h = 0.75$, $m_e = 0.067\ m_0$, and dielectric constant $\epsilon_1 = 13.1$. The units used here are the electron Rydberg energy $R_e = m_e/\epsilon_1^2\ R_a$, where R_a is the atomic Rydberg energy, for the energy and the atomic Bohr radius a_0 for the length scale. The two-dimensional exciton diameter is $d_\mu = 419\ a_0$ and the two-dimensional exciton Rydberg is $R_\mu = 0.4287\ R_e$. The "small parameter" λ here is running up to ten. As it is to be expected, the kinetic energy, i.e. the zeroth-order perturbation theory contribution dominates for $\lambda < 1$. Surprisingly, the predictions of the third-order perturbational calculation are very close to the results obtained through variational calculations [Bryant (1988)], especially for $\lambda < 5$, which at their turn describe correctly also the two-dimensional bulk limit ($-4R_\mu = -1.7148\ R_e$) for $\lambda \to \infty$. However, this feature is not preserved quite as well for smaller electron-hole mass ratios.

For the molecular (biexciton) binding energy, including also third order corrections one finds again an enhancement by a factor of up to 8, depending on the electron-hole mass ratio, as compared to the two-dimensional bulk value [Kleinmann (1988)]. The numerical results are illustrated in Fig. 3.2.

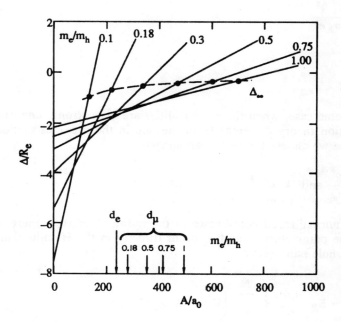

Fig. 3.2 The binding energy of the biexciton in a two-dimensional box within third order perturbation theory [Bryant (1990)] as a function of the lateral size A in nanometers (nm) for various electron–hole mass ratios.

3-2. Intermediate Confinement

For the case when the effective mass of the holes is much bigger than that of the electrons ($m_e/m_h \ll 1$) one may encounter [Efros and Efros (1982)] a peculiar situation where the radius of the microsphere is small relative to the electron Bohr radius

$$a_e = \frac{\epsilon_1 \hbar^2}{m_e e^2}, \tag{3.34}$$

but still big in comparison to the Bohr radius of the hole

$$a_h = \frac{\epsilon_1 \hbar^2}{m_h e^2} ,$$

(3.35)

so that

$$a_h < R < a_e .$$

(3.36)

In the extreme case, when these inequalities are very strong, one can apply a perturbation theory different from the one in the previous section. For this purpose we choose the small parameters

$$\lambda_1 = \frac{R}{a_e} \quad \text{and} \quad \lambda_2 = \frac{a_h}{R} .$$

(3.37)

and again normalize all coordinates to the radius R of the sphere. Consequently, the parameters λ_1 and λ_2 occur directly in the Hamiltonian of the N electron–hole pair system,

$$\mathcal{H}^{(N)} = E_R \frac{a_B a_e}{R^2} \left\{ \sum_{i=1}^{N} \left[\nabla_{e_i}^2 + \lambda_1 \lambda_2 \nabla_{h_i}^2 \right] \right.$$

$$+ \lambda_1 \sum_{\substack{i,j=1 \\ i \neq j}}^{N} \left(\frac{1}{|\mathbf{x}_{e_i} - \mathbf{x}_{e_j}|} + \frac{1}{|\mathbf{x}_{h_i} - \mathbf{x}_{h_j}|} \right) - 2\lambda_1 \sum_{i,j=1}^{N} \frac{1}{|\mathbf{x}_{e_i} - \mathbf{x}_{h_j}|} \right\}.$$

(3.38)

In order to simplify the discussion, we ignore in this chapter the difference between the dielectric constants of the dot and the surrounding medium. The boundary conditions again are

$$\Psi(\mathbf{x}_{e_i}, \mathbf{x}_{h_i}) \Big|_{x_i = 1} = 0 \quad \text{for all } i = 1, 2, ..., N.$$

(3.39)

Leaving aside the general energy scaling factor

$$E_R \; \frac{a_B \, a_e}{R^2} \; ,$$

we have the following structure of the scaled Hamiltonian,

$$\mathcal{H} = T_e + \lambda_1 \lambda_2 \, T_h + \lambda_1 \, W , \tag{3.40}$$

where the terms on the RHS correspond to the kinetic energy of the electrons, the kinetic energy of the holes, and the Coulomb interaction energy, respectively.

Next, we apply first-order perturbation theory with respect to the small parameter λ_1. The peculiarity of this case consists in the fact that the unperturbed Hamiltonian,

$$\mathcal{H}_0 = T_e , \tag{3.41}$$

does not depend on the hole coordinates, and therefore the unperturbed ground-state energy E_0 coincides with that of the non-interacting electron system. Hence, the unperturbed ground state has an infinite degeneracy with respect to the hole degrees of freedom. It is the product of the ground state of the non-interacting electrons with an arbitrarily normalized state for the holes,

$$|0\rangle = |\Psi_e^0\rangle \, |\psi_h\rangle \quad \text{and} \quad T_e \, |\Psi_e^0\rangle = E_0 \, |\Psi_e^0\rangle \tag{3.42}$$

First-order perturbation theory breaks the degeneracy. As usual in such situations, one has to minimize the first-order energy correction

$$E_1 = \lambda_1 \, \langle 0| \, \lambda_2 \, T_h + W \, |0\rangle$$

$$= \lambda_1 \, \langle \psi_h | \, [\lambda_2 \, T_h + \langle \Psi_e^0| \, W \, |\Psi_e^0\rangle] \, |\psi_h\rangle \tag{3.43}$$

with respect to the degenerate unperturbed wavefunctions. These are the normalized, but yet undetermined zeroth-order wavefunctions of the holes. This approach is, however, just the variational formulation of the Schrödinger eigenvalue problem and therefore its solution is given by the lowest eigenfunction of the stationary Schrödinger equation

$$\left[\lambda_2 \, T_h + \langle \Psi_e^0| \, W \, |\Psi_e^0\rangle \right] |\psi_h\rangle = E_h \, |\psi_h\rangle , \tag{3.44}$$

and

$$E_1 = \lambda_1 E_h . \tag{3.45}$$

Equation (3.44) describes the motion of the holes in the averaged field of the electrons. The averaging is performed over the confined state of the independent electrons. This corresponds exactly to the treatment of intermediate quantum confinement as proposed by Efros and Efros (1982).

Let us examine in detail the equations for the cases $N = 1$ (exciton) and $N = 2$ (biexciton). For the biexciton we study the case where the electrons have opposite spins and are therefore not identical. The normalized ground-state wavefunction of a "free" particle in the sphere of unit radius is given as (see Chap. 2)

$$\psi_{0,0,1}(\mathbf{x}) = \frac{\sin(\pi x)}{x\sqrt{2\pi}} \qquad (x \equiv |\mathbf{x}|).$$

In the case of the exciton the average potential of Eq. (3.44) acting on the hole is then

$$V(x) = -2 \int d^3y \ |\psi_{0,0,1}(\mathbf{y})|^2 \ \frac{1}{|\mathbf{y} - \mathbf{x}|}$$

$$= -4 \int_0^1 dy \ \sin^2(\pi y) \left[\frac{\theta(x-y)}{x} + \frac{\theta(y-x)}{y} \right]. \tag{3.46}$$

In Fig. 3.3 we plot the numerical evaluation of Eq. (3.46). One clearly sees the smooth shape of the average potential.

In the case of the biexciton the average potential acting on the two holes is [Bányai *et al.* (1988a) and (1988b)]

$$V(\mathbf{x}_1, \mathbf{x}_2) = \frac{2}{|\mathbf{x}_1 - \mathbf{x}_2|} - 4 \int d^3y \; |\psi_{0,0,1}(\mathbf{y})|^2 \left[\frac{1}{|\mathbf{y} - \mathbf{x}_1|} - \frac{1}{|\mathbf{y} - \mathbf{x}_2|} \right]$$

$$+ 2 \int d^3y_1 \, d^3y_2 \; |\psi_{0,0,1}(\mathbf{y}_1) \, \psi_{0,0,1}(\mathbf{y}_2)|^2 \; \frac{1}{|\mathbf{y}_1 - \mathbf{y}_2|}.$$

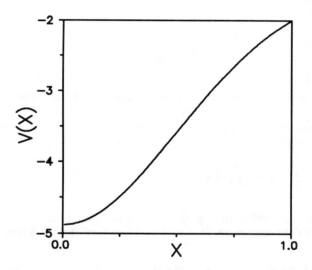

Fig. 3.3 The effective potential for the hole, averaged over the electron degree of freedom.

This potential can be written as

$$V(\mathbf{x}_1, \mathbf{x}_2) = \frac{2}{|\mathbf{x}_1 - \mathbf{x}_2|} + 2 \, V(x_1) + 2 \, V(x_2) + 3.572 \; . \tag{3.47}$$

A next simplifying step can be performed due to the smallness of the parameter λ_2. Actually, a small value of λ_2 corresponds to a heavy hole mass in Eq. (3.44). On the other hand, a heavy particle in a smooth potential well is most likely found close to the bottom of that potential. (This was not true for the Coulomb potential because it is not bounded from below, but it is true for the smooth averaged Coulomb potential.) Therefore, we get the estimate,

$$E_h \simeq \min \langle \Psi_e^0 | \ W \ | \Psi_e^0 \rangle \ . \tag{3.48}$$

This approximation should be understood as a lower bound for the energy E_h, since it ignores the positive contribution of the kinetic energy of the holes.

In the case of the exciton the minimum of the potential is at the center ($x = 0$) and its value for $x = 0$ is $- 4.876$ (see Fig. 3.3). If we recall the change of the energy scale, we see that our estimate for the exciton energy is

$$E_x \simeq \left(\frac{a_B \, a_e}{R^2} \pi^2 - 4.876 \, \frac{a_B}{R} \right) E_R \ . \tag{3.49}$$

In the case of the biexciton it is natural to look for the minimum of the potential in the configuration where the coordinates of the two holes are just mirror images ($\mathbf{x}_1 = - \mathbf{x}_2$). In this case the potential energy depends only on a single parameter $x = |\mathbf{x}_1| = |\mathbf{x}_2|$

$$V_b(x) = \frac{1}{x} + 4 \ V(x) + 3.572 \ . \tag{3.50}$$

This potential is plotted in Fig. 3.4. Its minimum is at $x = 0.28$, and the actual value at the minimum is $- 10.45$. Therefore the biexciton energy is

$$E_{xx} \simeq \left(\frac{a_B \, a_e}{R^2} 2\pi^2 - 10.45 \, \frac{a_B}{R} \right) E_R \ . \tag{3.51}$$

For the molecular binding energy we find

$$E_{mol} = 2 \ E_x - E_{xx} \simeq 0.698 \, \frac{a_B}{R} \, E_R \ , \tag{3.52}$$

again a strictly positive estimate. A numerical evaluation of the motion of the holes in the average potential of the electrons yields [Bányai et al. (1988b)]

$$E_{mol} \simeq \left[0.4 - 1.8 \sqrt{\frac{a_h}{R}} \right] \frac{a_B}{R} E_R , \qquad (3.53)$$

which is smaller, but still compatible with our previous estimates. It is however important to take into account that this last result holds only when $a_h/R \ll 1$. Therefore, it is questionable to predict on its ground a possible sign change of E_{mol}. These results contrast with other numerical estimates [Vandichev *et al.* (1987), Ekimov and Efros (1988)], where, within the same approach, a negative E_{mol} has been obtained. However, the detailed numerical calculations discussed in the following Chap. 4 show that the sign of E_{mol} remains indeed always positive for this model.

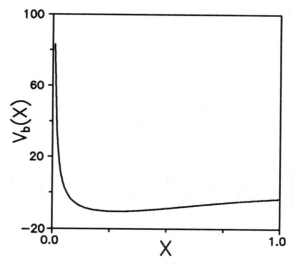

Fig. 3.4 The averaged potential for the holes in the biexciton problem, assuming mirror image positions of the holes, Eq. (3.50).

3-3. Weak Confinement

After the discussion of the strong and intermediate quantum confinement regimes in the preceding two sections, we now treat the case of large quantum dots ($R \gg a_e$, a_h). The confinement effects in this size regime are relatively small. Since the limit of an infinitely big sphere (bulk material) is well-known it seems at first that one should be able to treat the

effect of the confinement in the weak confinement case as a small perturbation of the bulk limit. In what follows, however, we will show that this
seemingly harmless limit is actually a very difficult one for a correct mathematical treatment, but at the same time it yields a substantial amount of
interesting physics.

We introduce the small parameter

$$\lambda = \frac{a_B}{R} \ll 1 \, , \tag{3.54}$$

and, as usual, we rescale the coordinates to the sphere radius R,

$$\mathbf{r} = \mathbf{x} \, R \, .$$

The Hamiltonian becomes

$$\mathcal{H}^{(N)}(\mathbf{r}) \ = \ \lambda \, E_R \, \mathcal{H}^{(N)}(\mathbf{x}) \, , \tag{3.55}$$

where

$$\mathcal{H}^{(N)}(\mathbf{x}) = - \lambda \sum_{i=1}^{N} \left(\frac{m_r}{m_e} \nabla_{e_i}^2 + \frac{m_r}{m_h} \nabla_{h_i}^2 \right)$$

$$+ \sum_{\substack{i,j=1 \\ (i \neq j)}}^{N} \left[\frac{1}{|\mathbf{x}_{e_i} - \mathbf{x}_{e_j}|} + \frac{1}{|\mathbf{x}_{h_i} - \mathbf{x}_{h_j}|} \right] - 2 \sum_{i,j=1}^{N} \frac{1}{|\mathbf{x}_{e_i} - \mathbf{x}_{h_j}|} \, . \tag{3.56}$$

In Eq. (3.56) we ignored the dielectric polarization effects due to the boundary, since they are expected to play a less important role for the case of
relatively big spheres. The wavefunctions obey the usual boundary conditions on the surface of the unit sphere.

The main purpose of rescaling is to put into evidence that the small
parameter λ now occurs as prefactor of the kinetic energy, similar to the
case of the intermediate confinement regime discussed in the previous Sec.
3-2. There, the smallness of the kinetic energy was due to the large hole
mass and we could argue that in the limit of infinite mass the hole would
be at the minimum of the smooth effective potential created by the electron. Now, however, due to the singular Coulomb potential all particles of
opposite sign would collapse to the same point within this approximation.

This is certainly incorrect, since we already know that for $\lambda = 0$ we have to recover the known bulk results.

The situation can be understood better by noting that a small λ may be interpreted either as a heavy mass or as a small Planck constant $\hbar$. Indeed, λ plays the role of $\hbar^2$. Therefore, we are faced with the problem of the quasiclassical approximation for the Coulomb motion of particles enclosed in a sphere. Unfortunately, this is still an unsolved mathematical problem.

We recall, that in the formal theory of the quasiclassical approximation, one has a McLaurin expansion of the phase of the wavefunction,

$$\psi = \exp\left[\frac{i}{\hbar} S \right],$$

which starts with the singular term

$$S_0/\hbar ,$$

where S_0 is the classical Hamilton-Jacobi action. Therefore, we may expect at best an asymptotic expansion for the eigenvalues and not a convergent Taylor expansion.

To illustrate the problem better let us consider the exactly solvable case of the motion of a particle inside the sphere which is subject to the attractive Coulomb potential of an external charge placed at the center of the sphere,

$$\left[-\frac{\hbar^2}{2m_r} \nabla^2 - \frac{e^2}{r} - E \right] \psi(\mathbf{r}) = 0 . \tag{3.57}$$

For the s-wave

$$\psi_0(r) = \frac{u_0(r)}{r}$$

we get the one-dimensional Schrödinger equation

$$\left[-\frac{d^2}{d\rho^2} - \frac{2}{\rho} - \tilde{E} \right] u_0(\rho) = 0 , \tag{3.58}$$

where we used the notations

$$\rho = \frac{r}{a_B} \quad \text{and} \quad \tilde{E} = \frac{E}{E_R} .$$

The boundary conditions are

$$u_0(0) = 0 \quad \text{and} \quad u_0\left(\frac{1}{\lambda}\right) = 0 . \tag{3.59}$$

These conditions assure the finiteness of the wavefunction ψ at the origin and infinity.

Equation (3.58) has two independent solutions for negative $\tilde{E}$, given by the two Whittaker functions

$$W_{|\tilde{E}|^{-1/2}, \, 1/2} (2\rho|\tilde{E}|^{1/2}) \quad \text{and} \quad W_{-|\tilde{E}|^{-1/2}, \, 1/2} (-2\rho|\tilde{E}|^{1/2}) .$$

The function u_0 is a linear combination of these Whittaker functions. From the two boundary conditions we obtain the transcendental equation for the eigenvalues,

$$W_{|\tilde{E}|^{-1/2}, \, 1/2} (0) \, W_{-|\tilde{E}|^{-1/2}, \, 1/2} \left(-\frac{2}{\lambda} |\tilde{E}|^{1/2}\right)$$

$$= W_{-|\tilde{E}|^{-1/2}, \, 1/2} (0) \, W_{|\tilde{E}|^{-1/2}, \, 1/2} \left(\frac{2}{\lambda} |\tilde{E}|^{1/2}\right) . \tag{3.60}$$

Using the known asymptotic behavior of the Whittaker functions,

$$W_{\nu,\mu}(z) \simeq z^\nu \, e^{-z/2} \quad \text{for} \quad |\arg z| < \pi , \tag{3.61}$$

we see that indeed the ground-state solution for $\lambda \to 0$ is $\tilde{E} \to 1$, and that the corrections are exponentially small and negative, i.e.

$$\tilde{E} \simeq 1 - \mathcal{O}(e^{-2/\lambda}) . \tag{3.62}$$

At first glance this situation might seem an simple one, but unfortunately this is not the case. Actually, it shows only that an expansion in powers of the small parameter λ is not possible because the energy is not analytical in λ around the origin. In the case of N Coulomb interacting particles in the sphere, it is hard to draw any definite conclusion regarding the asymptotic behavior in λ.

Recently Stolz (1989) has proved an interesting theorem relevant for the ground state of an electron-hole pair inside a microsphere. Here we only present his result using physicists terminology, without even trying to describe its complicated mathematical proof.

Theorem: *If the ground state of the Coulomb system in bulk is a bound state with the energy $E_0(\infty)$, then the ground-state energy $E_0(R)$ of the system in the sphere of radius R differs from $E_0(\infty)$ by the confined kinetic energy of the center-of-mass motion, up to terms of order $\ln(R)/R^3$.*

In the form of an equation this theorem is

$$E_0(R) = E_0(\infty) + \frac{\hbar^2}{2N(m_e + m_h)} \left[\frac{\pi}{R}\right]^2 + \mathcal{O}\left[\frac{\ln R}{R^3}\right]. \tag{3.63}$$

This mathematical result can be viewed as the justification of the intuitive physical argument of Efros and Efros (1982), but only for the ground state of the system. Therefore, it seems a reasonable approximation for small λ to use the approximate electron-hole-pair wavefunction

$$\Psi_{n\ell m}(\mathbf{x}_e, \mathbf{x}_h) = \phi(\mathbf{x})\, \psi_{n\ell m}(\mathbf{X}). \tag{3.64}$$

Here, the function

$$\phi(\mathbf{x}) = \frac{1}{\sqrt{\pi a_0^3}}\, e^{-x/a_0}, \tag{3.65}$$

with $\mathbf{x} = \mathbf{x}_e - \mathbf{x}_h$ describes the relative motion in the lowest bound state of the bulk material, whereas $\psi_{n\ell m}(\mathbf{X})$ is the wavefunction for the confined motion of the mass center

$$\mathbf{X} = \frac{m_e\, \mathbf{x}_e + m_h\, \mathbf{x}_h}{m_e + m_h}, \tag{3.66}$$

$$\psi_{n\ell m}(\mathbf{X}) = Y_{\ell m}(\theta, \phi)\, \sqrt{\frac{2}{R^3}}\, \frac{j_\ell(\kappa_{n,\ell}\, X/R)}{j_{\ell+1}(\kappa_{n,\ell})}, \tag{3.67}$$

see Chap. 2.

The wavefunction (3.64) is an exact solution of the one electron-hole pair stationary Schrödinger equation (Wannier equation), corresponding to the exciton eigenenergies

$$\epsilon_{n\ell} = E_G - E_R + \frac{\hbar^2 \, \kappa_{n,\ell}^2}{2MR^2} , \tag{3.68}$$

where $M = m_e + m_h$ is the exciton mass. However, the wavefunctions (3.67) do not exactly satisfy the boundary condition, i.e., they do not vanish at the surface of the dot. This violation occurs in a spherical shell of width a_B which is small in comparison with R. Hence, one hopes that the errors introduced by this approximation are also small. An analogous approximation can be obtained also for the two-pair states, leading to the biexciton in the bulk.

The discussed approximation scheme may be expressed in the formalism of second quantization using the Bose operators $b_{n\ell m}^\dagger$ and $b_{n\ell m}$ for electron-hole-pair creation and annihilation, respectively. The Hamiltonian of electron-hole pairs in this model is given by

$$\mathcal{H} = \sum_{n,\ell,m} \left[E_G - E_R + \frac{\hbar^2 \, \kappa_{\ell n}^2}{2MR^2} \right] b_{n\ell m}^\dagger \, b_{n\ell m} , \tag{3.69}$$

where

$$[b_{n\ell m} , b_{n'\ell'm'}^\dagger] _- = b_{n\ell m} \, b_{n'\ell'm'}^\dagger - b_{n'\ell'm'}^\dagger \, b_{n\ell m}$$

$$= \delta_{\ell,\ell'} \, \delta_{m,m'} \, \delta_{n,n'}$$

$$[b_{n\ell m} , b_{n'\ell'm'}] _- = 0 = [b_{n\ell m}^\dagger , b_{n'\ell'm'}^\dagger] _- . \tag{3.70}$$

The Hamiltonian (3.69) resembles the elementary model of Chap. 2 describing uncorrelated electrons and holes, but here the quantum numbers $\ell m n$ refer to the motion of the mass center of a pair. Consequently, also the expression for the interband polarization operator in terms of the new Bose creation and annihilation operators is different from that in Chap. 2. In fact, the first part of Eq. (2.42), which is also correct here, predicts that the matrix element of the interband polarization operator between the vacuum and the electron-hole pair state is proportional to the integral of the exciton wavefunction with the electron and hole coordinates being equal

$$\langle 0|P| \text{ exciton} \rangle = \int d^3x \ \Psi_x(\mathbf{x},\mathbf{x}) \ . \tag{3.71}$$

Using Eq. (3.64) this can be written as

$$\langle 0|P|n\ell m \rangle = \int d^3x \ \Psi_{n\ell m}(\mathbf{x},\mathbf{x}) = \delta_{\ell,0} \ \delta_{m,0} \ \frac{1}{n} \ \sqrt{2\left(\frac{R}{\pi a_0}\right)^3} \ , \tag{3.72}$$

and consequently

$$P = p_{cv} \ \sqrt{2} \ \left(\frac{R}{\pi a_0}\right)^{3/2} \sum_n \frac{1}{n} \ (b_{n00} + b_{n00}^\dagger) \ . \tag{3.73}$$

The corresponding optical susceptibility of a dot is then

$$\chi(\omega) = -\frac{3}{2\pi} \frac{|p_{cv}|^2}{(\pi a_B)^3} \sum_n \frac{1}{n^2} \frac{1}{\hbar\omega - E_G + E_R - \dfrac{\hbar^2 \ \kappa_{0n}^2}{2MR^2} + i\gamma} + (\omega \rightarrow -\omega) \tag{3.74}$$

Eq. (3.74) indicates the presence of infinitely many excitonic lines whose oscillator strengths decrease as n^{-2}.

A typical feature of the weak confinement configuration is that we get a large number of states with very small energy separations and, due to the lack of translational invariance, there is no momentum conservation that would forbid certain transitions between these states. Since each one of these levels has a homogeneous width, and since there is additional in-homogeneous broadening caused by the size distribution of different microspheres, the energy levels overlap, leading to an effectively continuous spectrum. The situation is therefore quite different from the strong confinement regime, where the distances between the discrete confinement levels are so large, that one has a basically discrete spectrum. Consequently, it is not very useful to compute a few low lying levels with high precision. It is much more important to predict certain global characteristics such as the number of levels in a given energy interval and their average oscillator strength. As always in the treatment of systems with many degrees of freedom, the increased complexity brings also simplifications.

Chapter 4
ELECTRON-HOLE-PAIR STATES

In this chapter we present a variety of numerical schemes which were developed in the past few years to compute the energies and wavefunctions of electron-hole systems mainly in spherical semiconductor quantum dots. These methods deal with the electron-hole Coulomb interaction in a non-perturbative manner and are applicable for a wide range of parameters (radii, dielectric constants, barrier heights and effective masses). Although all the numerical methods are essentially variational in nature, we shall distinguish between two categories : *i*) methods, in which an analytical variational wavefunction is optimized with regard to some parameters; *ii*) methods in which the variational test function belongs to a linear subspace of the Hilbert space, whose dimension may be increased. The first ones are *intrinsically variational*, while the second ones are often called *numerically exact* or *configuration interaction* methods.

The main difference between the intrinsically variational and the configuration interaction methods lies in the convergent iterative nature of the configuration interaction approach. Increasing the dimension of the linear subspace allows reliable numerical checks of the convergence. Even though the intrinsically variational methods offer few possibilities to check their error, they are still very useful in providing compact analytical expressions for further applications. The validity of the intrinsically variational results should be checked carefully, e.g., through comparison with the numerically exact results.

In this chapter we first discuss a few examples of intrinsically variational calculations for the energetically lowest electron-hole-pair states in quantum dots (Sec. 4-1). The effects of a finite confinement potential are investigated and a simple model to estimate the influence of the quantum-dot environment is analyzed. In Sec. 4-2 we review the configuration interaction (matrix diagonalization) technique. The energies and wavefunctions for the one- and two-electron-hole-pair states in quantum dots are calculated. The biexciton binding energy is found to increase with decreasing dot size, demonstrating the important role of Coulomb correlation effects in small dots.

The quantum Monte Carlo technique is discussed in Sec. 4-3 as another numerical approach to calculate the electron-hole-pair states. The results of this method yield an independent check of the matrix diagonalization calculations of Sec. 4-2. Furthermore, the quantum Monte Carlo approach allows to compute system properties at finite temperatures, showing interesting temperature-dependent modifications of electron and hole distributions in the dot.

In Sec. 4-4 we analyze the influence of the induced surface polarization on the electron-hole-pair states. For finite confinement potential and different dielectric constants inside and outside of the dot, we find a transition of the electron-hole-pair state from a "volume state" to a "surface trapped state". In the volume state, electron and hole are localized inside the dot, similar to the situation with an infinite confinement potential. For smaller dots or weaker quantum confinement potential the surface trapped state is realized where the hole is localized in a shell just outside the quantum dot.

The confinement-induced mixing of the valence states causes modifications of the allowed hole energies in a quantum dot. A brief discussion of these phenomena is presented in Sec. 4-5. It is shown that the quantum confinement removes degeneracies of the bulk energy bands and leads to substantial mixing of the hole states with different angular momentum.

In the last section of this chapter (Sec. 4-6) we review calculations based on a lattice model of quantum dots. The results obtained can be regarded as a justification of the effective mass approximation.

4-1. Variational Calculations

In this section we discuss a few examples of intrinsically variational calculations. A very successful proposal for a variational test function for the ground state motion of an electron-hole pair in a sphere with rigid barriers without the surface dielectric polarization corrections is [Kayanuma (1986)]

$$\psi(\mathbf{r}_e, \mathbf{r}_h) = e^{-\lambda_s |\mathbf{r}_e - \mathbf{r}_h|} \, \phi_{e,s}(r_e) \, \phi_{h,s}(r_h) \,. \tag{4.1}$$

Here λ_s is the variational parameter and $\phi_{i,s}(r_i)$ is the one-particle ground-state wavefunction in the infinite spherical potential well. For $R \to \infty$ and $\lambda_s \to 1/a_B$, where a_B is the bulk-exciton Bohr radius, the variational wavefunction (4.1) approaches the hydrogen $1s$-wavefunction for the relative motion of the electron-hole pair. At the same time, Eq. (4.1) yields the proper strong confinement limit for $R/a_B \ll 1$.

Even though more elaborate wavefunctions with more variational parameters may be constructed [Nair at *al* (1987)], we restrict our discussion to the simplest and most transparent case. Of special interest here is the extension of the variational calculations to analyze quantum dots with finite potential barriers [Tran Thoai *et al.* 1990]. For this purpose one starts again with the same ansatz, Eq. (4.1), but the one-particle wavefunctions ϕ are now solutions of the motion of a single particle in a finite spherical potenial well. Furthermore, one might take into account that the effective masses of the particles inside and outside the well may be different. This last point can be very important since the semiconductor material of the dot often differs substantially from that of its surrounding.

We write the confinement potential in the form

$$U_i(r_i) = \begin{cases} 0 & \text{for} \quad r_i \leq R \\ U_{oi} & \text{for} \quad r_i \geq R \end{cases} . \tag{4.2}$$

The functions ϕ_e (ϕ_h) for the electron (hole) in Eq. (4.1) are the radial eigenfunctions corresponding to the lowest energy state in the spherical well given by Eq. (4.2). These wavefunctions can be written as

$$\phi_{i,s}(r_i) = \frac{\sin (k_{i,s} r_i)}{k_{i,s} r_i} \tag{4.3a}$$

for $r_i \leq R$ and

$$\phi_{i,s}(r_i) = \frac{\sin (k_{i,s} R)}{k_{i,s} r_i} e^{-q_{i,s}(r_i - R)} \tag{4.3b}$$

for $r_i \geq R$ and $i = e, h$. Here we introduced

$$k_{i,s}^2 = \frac{2m_{i,1} E_{i,s}}{\hbar^2} , \tag{4.4}$$

$$q_{i,s}^2 = \frac{2m_{i,2} U_{oi}}{\hbar^2} - k_{i,s}^2 \frac{m_{i,2}}{m_{i,1}} , \tag{4.5}$$

and $m_{e,1}$, $m_{h,1}$, and $m_{e,2}$, $m_{h,2}$ denote the effective mass of the electron and hole inside and outside the microcrystallite.

From the boundary conditions that $\phi_i(r_i)$ and $1/m_i \ \partial\phi_i(r_i)/\partial r_i$ are both continuous at $r_i = R$ we obtain

$$\tan(k_{i,s}R) = \frac{k_{i,s}R}{1 - (1 + q_{i,s}R) \, m_{i,1}/m_{i,2}} \ , \quad i = e, h \ , \tag{4.6}$$

which determines the ground-state energy $E_{e,s}$ ($E_{h,s}$) of the noninteracting electron (hole) in the potential well.

The ground-state energy E_s of the interacting electron-hole pair is computed from the expectation value of the Hamiltonian

$$\mathcal{H}_x = - \frac{\hbar^2 \nabla_e^2}{2m_e} - \frac{\hbar^2 \nabla_h^2}{2m_h} - \frac{e^2}{\epsilon |\mathbf{r}_e - \mathbf{r}_h|} + U_e(r_e) + U_h(r_h) \tag{4.7}$$

by minimizing the expression

$$E_s = \int d\mathbf{r}_e \int d\mathbf{r}_h \ \psi(\mathbf{r}_e, \mathbf{r}_h) \ \mathcal{H}_x \ \psi(\mathbf{r}_e, \mathbf{r}_h) \tag{4.8}$$

with respect to the variational parameter λ_s.

To compute the first excited state a variational wavefunction is needed which exhibits the appropriate limiting bahavior. This wavefunction should reduce to the strong-confinement product of p-state functions for the electron and hole (for $R \to 0$) and it should reproduce the $2s$-wavefunction for the electron-hole relative motion (for large quantum dots). For this purpose Tran Thoai et al. (1990) choose

$$\Phi(\mathbf{r}_e, \mathbf{r}_h) = \phi_{e,p}(r_e) \ \phi_{h,p}(r_h) \cos(\gamma) \left[(\delta - \lambda_p |\mathbf{r}_e - \mathbf{r}_h|) \ e^{-\lambda_p |\mathbf{r}_e - \mathbf{r}_h|} \right] \ , \tag{4.9}$$

where

$$\phi_{i,p}(r_i) = \frac{\sin(k_{i,p} r_i)}{k_{i,p}^2 r_i^2} - \frac{\cos(k_{i,p} r_i)}{k_{i,p} r_i} \tag{4.10a}$$

for $r_i \le R$ and

$$\phi_{i,p}(r_i) = \frac{[\sin(k_{i,p}R) - k_{i,p}R\cos(k_{i,p}R)]q_{i,p}^2}{(q_{i,p}R + 1)\,k_{i,p}^2}\;\frac{e^{-q_{i,p}(r_i - R)}}{q_{i,p}r_i}\left[\frac{1}{q_{i,p}r_i} + 1\right]$$

$$(4.10b)$$

for $r_i \geq R$. The parameter γ in Eq. (4.9) denotes the angle between $\mathbf{r}_e$ and $\mathbf{r}_h$, and δ and λ_p are the variational parameters. One of these parameters is fixed by the condition that the wavefunction (4.9) has to be orthogonal to the ground-state wavefunction (4.1).

As for the ground-state calculation, the boundary conditions for the wavefunction are used to determine the excited-state energies of the non-interacting electrons and holes. One obtains the transcendental relation

$$\tan(k_{i,p}R) = \frac{k_{i,p}R}{1 + \dfrac{(k_{i,p}R)^2}{\dfrac{m_{i,1}}{m_{i,2}}\dfrac{2 + 2q_{i,p}R + q_{i,p}^2 R^2}{1 + q_{i,p}R} - 2}}\,, \qquad (4.11)$$

where $k_{i,p}$ and $q_{i,p}$ are defined by Eqs. (4.4) and (4.5) with the index s replaced by p. The expectation value of the quantum dot Hamiltonian (4.7) with the wavefunction (4.9) gives the excited-state energy E_p of the interacting electron-hole pair.

Tran Thoai *et al.* (1990) minimize the energies and E_s and E_p numerically by variation of λ_s and λ_p, respectively. The calculations were done for material parameters appropriate for CdS, i.e. effective electron mass inside the quantum dot $m_{e,1} = 0.235\, m_0$, effective hole mass $m_{h,1} = 1.35\, m_0$, band gap $E_g = 2.583$ eV, exciton binding energy $E_R = 27$ meV, and exciton Bohr radius $a_B = 30.1$ Å For simplicity, tunneling effects of the hole were neglected by choosing $U_{oh} = \infty$, i.e. the hole is confined inside the microcrystallites.

Fig. 4.1 shows the energies of the two quantum-confined states as function of quantum-dot radius for a confinement potential $U_{oe} = 40\, E_R$ and $m_{e,2} = m_0$. This situation is more or less appropriate for quantum dots in a liquid solution, where it is assumed that the electrons can move freely once they have tunneled out of the crystallites. The solid lines are obtained assuming an infinite confinement potential (ideal quantum confinement) and the dashed lines are the results of the variational calculations. As expected, the lower potential barrier reduces the quantum-confined energy levels. The resulting relative changes are larger for the excited state since the reduced confinement energy contribution for a finite confinement barrier is more pronounced for energetically higher states. For radii R smaller than $\simeq 0.4\, a_B$ the calculations predict the disappearance of an excited state that is bound inside the quantum dot. In the limit $R \gg a_B$ the results approach the correct bulk semiconductor values of $-E_R$ for the ground

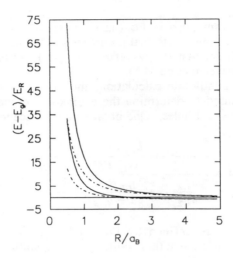

Fig. 4.1: Computed energies of the two lowest dipole-allowed quantum-confined states are plotted as function of crystallite radius. The solid lines show the results for an infinite confinement potential and the dashed lines are for a confinement potential $U_{oe} = 40\ E_R$. The other material parameters are $m_{e,1} = 0.235\ m_0$, $m_{e,2} = m_0$, $m_{h,1} = 1.35\ m_0$, $E_g = 2.583$ eV, $E_R = 27$ meV, and $a_B = 30.1$ Å From Tran Thoai et al. (1990).

state and $-E_R/4$ for the excited state, respectively.

To demonstrate the dependence of the energy states on the strength of the confinement potential we plot in Fig. 4.2 the computed energy levels as function of V_{oe} assuming a crystallite radius $R = 0.5\ a_B$ and otherwise the same conditions as in Fig. 4.1. Fig. 4.2 shows that both energy levels increase with increasing quantum confinement. The increase of the excited state is clearly more pronounced than that of the ground-state level. For a confinement potential which is less than $\simeq 30\ E_R$ the first excited state is no more bound inside the quantum dot for the parameters chosen.

Fig. 4.3 shows the dependence of the energy levels on the electron mass outside the semiconductor material for $R = 0.5\ a_B$ and $U_{oe} = 80\ E_R$. We see that the energies decrease with increasing outside mass. This indi-

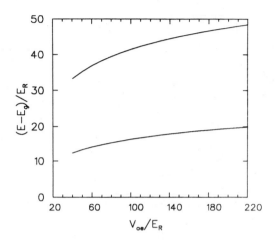

Fig. 4.2: Computed energies of the two lowest quantum confined states as function of the confinement potential for a quantum dot of radius $R = 0.5\ a_B$. The other parameters are the same as in Fig. 4.1. From Tran Thoai *et al.* (1990).

cates that for the example of quantum dots in glass, where the electrons outside the microcrystallites are most likely in localized states simulated by a very heavy mass, the confined energy levels are lower than those for the same size quantum dots in a liquid solution. The overall energy variation as function of mass ratio was verified by solving the Schrödinger equation for a one-dimensional potential well with different masses inside and outside, using the same boundary conditions discussed above, i.e., $\phi_i(r_i)$ and $1/m_i\ \partial\phi_i(r_i)/\partial r_i$ are continuous at $r_i = R$. It is always found that the energy of the confined states decreases with increasing outside mass.

Fig. 4.4 compares the computed ground-state energies for $U_{oe} = 40E_R$ (dashed line) and $U_{oe} = \infty$ (solid line) to experimental results for CdS colloids. The radius of the CdS crystallites was determined by electron microscopy (triangles) [Weller *et al.* (1986), Kunath *et al.* (1985)] or by the fluorescence quenching method (squares) [Weller *et al.* (1986), Ramsden and Graetzel (1984)]. The results computed for a finite potential barrier show good agreement with the experimental observations, whereas the infinite potential-barrier calculation clearly overestimates the confinement energy.

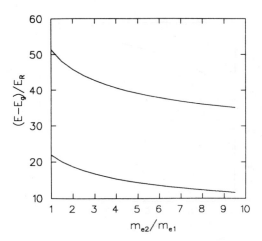

Fig. 4.3: Computed energies of the two lowest quantum confined states as function of the electron mass outside the quantum dot for $R = 0.5\,a_B$, $U_{oe} = 80\,E_R$ and otherwise the same parameters as in Fig. 4.1. From Tran Thoai *et al.* (1990).

A generalization of Kayanuma's variational approach was proposed also for the two-pair problem. Takagahara (1988) chooses the wavefunction as

$$\psi(\mathbf{r}_{e_1}, \mathbf{r}_{e_2}, \mathbf{r}_{h_1}, \mathbf{r}_{h_2}) \propto \prod_{\substack{\alpha=e,h \\ i=1,2}} \frac{\sin(\pi r_{\alpha_i}/R)}{r_{\alpha_i}} \exp\left[-\gamma_3\,r_{h_1,h_2}\right]$$

$$\times \left(\exp\left[-\gamma_1\left(r_{e_1,h_1} + r_{e_2,h_2}\right)\right] + \exp\left[-\gamma_2\left(r_{e_1,h_2} + r_{e_2,h_1}\right)\right]\right), \quad (4.12)$$

where

$$r_{\alpha_i,\beta_j} \equiv \left|\mathbf{r}_{\alpha_i} - \mathbf{r}_{\beta_j}\right| \quad (\alpha, \beta = e, h; \; i, j = 1, 2). \quad (4.13)$$

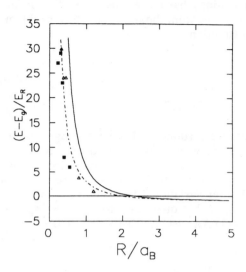

Fig. 4.4: Computed ground state energy for a well depth $U_{oe} = 40\, E_R$ (dashed line) and $U_{oe} = \infty$ (solid line) in comparison with the experimental results of Weller *et al.* (1986). The parameters are the same as in Fig. 4.1. From Tran Thoai *et al.* (1990).

Taking into account also the dielectric surface polarization effects in the hamiltonian the parameters γ_i were found numerically minimizing the ground state energy.

Using the variational wavefunction (4.12) Takagahara (1988) computed two-electron-hole-pair energies and also molecular (biexciton) binding energies. For some conditions these binding energies exhibit a sign change for very small radii R, although the exact result for this model is a positive binding energy (see Chapter 3).

Actually, it is not too surprising that variational methods of the first kind might not give accurate evaluations of ground-state energy differences of two different Hamiltonian problems, like in the calculation of the molecular binding energy (twice the one-pair minus the two-pair ground state energies). A variational approximation can only give an upper bound of the ground-state energy for each Hamiltonian problem. Nothing is known about the energy differences of two independent approximations for two independent problems. Hence, one should use variational calculations of binding energies only with great care.

Variational calculations have been performed also for square shaped quasi-two-dimensional quantum boxes. For this problem Bryant (1988, 1989a) uses the trial function

$$\psi(\mathbf{x}_e, \mathbf{x}_h) \propto \prod_{\alpha = e,h \,;\, i = 1,2} \cos(\pi x_i^\alpha / A) \sum_n c_n \, e^{-\gamma_n |\mathbf{x}_e - \mathbf{x}_h|} . \quad (4.14)$$

The numerical results were found to be in good agreement with perturbational result for a relatively wide range of quantum dot sizes. Somewhat worse agreement was obtained for very small electron–hole mass ratios, since here the variational function has to take care of the strong electron–hole asymmetry and other variational test functions have to be considered.

For relatively large cubic quantum boxes ($A > 3a_B$) a more sophisticated trial function is [D'Andrea and Del Sole (1990)]

$$\psi_{i_1 i_2 i_3}(\mathbf{x}, \mathbf{X}) \propto \prod_{i = 1, 2, 3} R(x_i, X_i) \, e^{-x/a_B} \quad , \quad\quad\quad (4.15)$$

where $\mathbf{x}$ and $\mathbf{X}$ are the relative and center of mass coordinates respectively. The so called "recycling function" R is defined as

$$R(x, X) = \cos(KX) - F_e(x) \cosh(PX) + F_0(x) \sinh(PX) \quad , \quad\quad (4.16)$$

where

$$F_e(x) = \frac{\sinh(PX_1) \cos(KX_2) - \sinh(PX_2) \cos(KX_1)}{\sinh[P(X_1 - X_2)]} \quad , \quad\quad (4.17a)$$

$$F_0(x) = \frac{\cosh(PX_1) \cos(KX_2) - \cosh(PX_2) \cos(KX_1)}{\sinh[P(X_1 - X_2)]} \quad , \quad\quad (4.17b)$$

and

$$X_1 = \frac{L}{2} - x \, \frac{m_h}{M} \quad ; \quad X_2 = -\frac{L}{2} + x \, \frac{m_e}{M} \, .$$

The trial function (4.15) was used to show that the lowest exciton energies may be parametrized as

$$E_x \simeq - E_R + n^2 \, \frac{\hbar^2}{2M} \left[\frac{\pi}{L - 2/P} \right]^2 \quad (n = 1, 2, 3) \, .$$ (4.18)

The quantity $1/P$ defines an effective "dead layer" or "transition layer" for the exciton and its extension is, as can be expected on intuitive grounds, approximately equal to the exciton Bohr radius a_B. The appearance of this layer reflects the fact that the mass center of the exciton has to keep a distance from the quantum dot surface which is of the order of a_B. Consequently, the radius of the effective confinement volume is somewhat reduced, instead of R it is only $R - a_B$.

4-2. Matrix Diagonalization Technique

It is important for the success of variational calculations such as those discussed in the previous section 4-1 that the choice of the trial functions is guided by an educated guess. The quality of the results strongly depends on the wavefunctions chosen, and there is no systematic way to improve or check the accuracy of the results.

In this section we discuss an example of a "numerically exact" method, also called configuration interaction method, which allows to compute the energies and wavefunctions of one- and two-electron-hole-pair states in quantum dots [Hu et al. (1990a), (1990b), (1990c), Koch (1990)]. The main purpose of the calculations is to find numerical solutions with well defined precision for the energies and wavefunctions of the lowest one-pair and two-pair states. Such calculations are performed mainly for microspheres with radii R comparable to the exciton Bohr radius a_B (0.25 < R/a_B < 5). The method includes the treatment of surface dielectric effects and also the effects of finite confinement barriers. It is easily possible to vary parameters such as the dielectric constants or the electron-hole mass ratios.

The variational reservoir is taken from linear combinations of a finite set of eigenfunctions of the kinetic energy, which form a linear subspace of the Hilbert space to which the exact wavefunction belongs. Furthermore, the eigenfunctions are chosen to satisfy the correct boundary conditions, such as vanishing on the quantum dot surface for perfect confinement conditions or at infinity for a finite confinement potential. The minimization of the energy on this linear subspace is equivalent to the diagonalization of the finite Hamiltonian matrix taken on the considered set of orthonormalized eigenstates of the kinetic energy. Since the whole set of eigenstates of the kinetic energy forms a complete basis in the Hilbert space, the procedure of increasing the number of linearly independent eigenstates is converging to the exact result. The limits are set only by the capacity of the computer to diagonalize $N \times N$ hermitean matrices. One additional criterion for the proper choice of the eigenfunctions is to define the variational set as belonging to all unperturbed eigenvalues below a certain energy and then to increase this upper energy when improving the calculations.

We discuss the method here for the one electron-hole-pair problem in a rigid well [Hu (1991)]. For this purpose, we write the Schrödinger equation inside the dot as

$$(\mathscr{H}_x - E_x) \, \phi^1(\mathbf{r}_e, \mathbf{r}_h) = 0, \tag{4.19}$$

where $\mathscr{H}_x$ is the one-electron-hole-pair Hamiltonian

$$\mathscr{H}_x = \left[-\frac{1}{2m_e} \nabla_e^2 - \frac{1}{2m_h} \nabla_h^2 + W(\mathbf{r}_e, \mathbf{r}_h) \right], \tag{4.20}$$

see Eqs. (3.18) with (3.2) – (3.4).

As described in Sec. 2-5, the effective Coulomb energy W in the dot differs from the Coulomb potential

$$\frac{e^2}{\epsilon_1 |\mathbf{r}_1 - \mathbf{r}_2|}$$

due to surface polarization effects,

$$\delta W(\mathbf{r}_1, \mathbf{r}_2) = Q_1(r_1) + Q_1(r_2) - Q_2(\mathbf{r}_1, \mathbf{r}_2). \tag{4.21}$$

Here we write the contribution due to the interaction of the electron (or the hole) with its own induced surface charge as

$$Q_1(r) = \frac{e^2}{2R} \sum_{\ell=0}^{\infty} \alpha_\ell \left(\frac{r}{R} \right)^{2\ell}. \tag{4.22}$$

The surface-charge induced electron-hole interaction is

$$Q_2(\mathbf{r}_1, \mathbf{r}_2) = \frac{e^2}{R} \sum_{\ell=0}^{\infty} \alpha_\ell \left(\frac{r_1 r_2}{R} \right)^{\ell} P_\ell(\cos\theta). \tag{4.23}$$

Here, expansions in Legendre polynomials were used and θ is the angle between $\mathbf{r}_1$ and $\mathbf{r}_2$. The corresponding Legendre expansion of the Coulomb potential is given by Eq. (2.85). Furthermore, we introduced the abbreviation

$$\alpha_\ell = \frac{(\epsilon - 1)\ell + 1}{\epsilon_1(\ell\epsilon + \ell + 1)},$$

(4.24)

and ϵ is again the relative dielectric constant of the two media defined in Eq. (2.61).

It is possible to substantially reduce the number of states which need to be considered explicitly by utilizing the relevant symmetry considerations. For this purpose it is helpful to apply arguments from group theory. The Hamiltonian (4.20) is invariant with respect to simultaneous three-dimensional rotations of the electron and hole around the center of the sphere. This is a very helpful invariance, even though it is reduced in comparison to the symmetry group of the kinetic energy, which is invariant with respect to independent rotations of each particle.

The exact wavefunctions can be classified according to the irreducible representations of the rotation group. If we denote by L and M as usual the total orbital angular momentum quantum numbers, then according to the Wigner-Eckart theorem we can write for a scalar Hamiltonian

$$\langle L,M \mid \mathcal{H}_x \mid L',M' \rangle = \delta_{L,L'} \cdot \delta_{M,M'} \, (L\mid \mathcal{H}_x \mid L) \,,$$

(4.25)

where $(L\mid \mathcal{H}_x \mid L)$ is the reduced matrix element [see e.g. Schiff (1968), Sec. 28]. Hence, the diagonalization problem reduces to independent diagonalizations of the subspaces with given L. Furthermore, since the interband polarization operator Eq. (2.42) is invariant with respect to rotations of all particles around the center, any electron–hole–pair state that can be excited from the ground state has angular momentum zero ($L = 0$, $M = 0$). This means that one may restrict the whole variational reservoir to the linear subspace of eigenstates of the kinetic energy with zero total angular momentum.

Electron–hole–pair states with total angular momentum quantum numbers L, M may be formed from the products of the one-particle kinetic energy eigenfunctions with angular momentum quantum numbers ℓ, m with the help of the Clebsch-Gordan coefficients $\langle \ell_1,m_1;\ell_2,m_2 \mid L,M \rangle$,

$$|n_e,n_h,\ell_e,\ell_h,L,M\rangle = \sum_{m_1,m_2} \langle \ell_e,m_e;\ell_h,m_h|L,M\rangle \ |n_e,\ell_e,m_e\rangle \ |n_h,\ell_h,m_h\rangle \ .$$

$$(4.26)$$

Besides the total angular momenum quanum numbers L and M these states are characterized also by the quantum numbers ℓ_e, ℓ_h, n_e, and n_h. They form a complete orthonormalized basis. We can write the $L = M = 0$ state as

$$|n_e,n_h,\ell,\ell,0,0\rangle = \sum_m \langle \ell,m;\ell,m|\ 0,0\rangle \ |n_e,\ell,m\rangle \ |n_h,\ell,-m\rangle \ , \qquad (4.27)$$

where we took already into account that only equal and oppositely oriented electron and hole angular momenta can give rise to zero total angular momentum. Yet another helpful property lies in the two-particle character of the effective Coulomb interaction W, wich enables us to reduce the matrix elements in the two-pair space to those in the one-pair space (see Chap. 3).

Systematically exploiting all the above mentioned symmetry properties and the standard integrals for the Legendre polynomials, one may reduce the original six-fold integrals in the calculation of the matrix elements to double integrals over the radial coordinates, which are left for numerical evaluation [Hu (1991)]. The remaining problem is to diagonalize the Hamiltonian matrix

$$\langle n_e,n_h,\ell,0,0 \ | \ \mathscr{H}_x \ |n'_e,n'_h,\ell',0,0\rangle$$

where the indices ℓ, n_e and n_h run over a finite set of values. The kinetic energy is purely diagonal in this basis. The non-diagonal matrix elements are provided only by the effective Coulomb interaction W between the electron and the hole. In the numerical procedure one selects the basis vectors starting from the energetically lowest state and increasing the number of states until the desired accuracy is achieved [Hu et al. (1990a, 1990b, 1990c)]. The dimension of the basis is only limited by the capability of the computer.

As an example of the results we plot in Fig. 4.5 the calculated ground-state energy for the exciton for two different electron-hole mass-ratios as function of the quantum-dot radius. This figure clearly shows the sharp energy increase for smaller dots expected from the R^{-2}-dependence of the

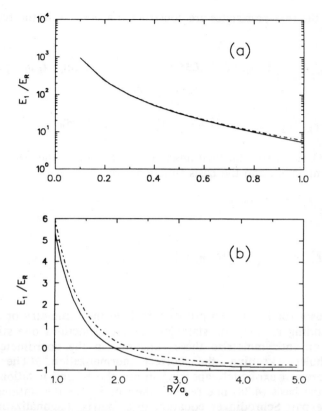

Fig. 4.5: The ground-state energy of the exciton for a mass ratio $m_e/m_h = 0.1$ (full line) respectively $m_e/m_h = 0.01$ (broken line). The pictures show the results for $R/a_B \leq 1$ (a) and $R/a_B \geq 1$ (b). From Hu *et al.* (1990b).

confinement energy. To demonstrate the dependence on the electron-hole mass ratio results for $m_e/m_h = 1$ and $m_e/m_h = 0.01$ are shown. The energy differences are most pronounced in the size regime $R/a_0 < 4$.

In order to apply the numerical matrix diagonalization method for the two-electron-hole-pair states, one is now confronted with the fact that four single-particle angular momenta have to be added properly. This could be done by adding the angular momenta of the already constructed one-pair angular momentum eigenstates of Eq. (4.26). It is however more convenient to add separately the angular momenta of the electrons and holes [Hu (1991)] in order to take into account the proper antisymmetrization for identical Fermions (after the corresponding consideration of the spin). Then the two-pair states are constructed according to the recipe

$$
|n_{e_1} n_{e_2} n_{h_1} n_{h_2} \ell_{e_1} \ell_{e_2} \ell_{h_1} \ell_{h_2}, \ell_e \ell_h, LM \rangle = \sum_{m_e m_h} \langle \ell_e, m_e; \ell_h, m_h | LM \rangle
$$

$$
\times |n_{e_1}, n_{e_2}, \ell_{e_1}, \ell_{e_2}, \ell_e, m_e \rangle |n_{h_1}, n_{h_2}, \ell_{h_1}, \ell_{h_2}, \ell_h, m_h \rangle , \tag{4.28}
$$

where (L, M) stand for the total angular momentum of the two-pair state and its z-component, and the basis functions are

$$
|n_{e_1}, n_{e_2}, \ell_{e_1}, \ell_{e_2}, \ell_e, m_e \rangle =
$$

$$
\sum_{m_{e_1}, m_{e_2}} \langle \ell_{e_1}, m_{e_1}; \ell_{e_2}, m_{e_2} | \ell_e m_e \rangle |n_{e_1}, \ell_{e_1}, m_{e_1} \rangle |n_{e_2}, \ell_{e_2}, m_{e_2} \rangle , \tag{4.29}
$$

and analogously for holes. To properly include the symmetry or antisymmetry (depending on the spin state) of the wavefunctions one still has to symmetrize or antisymmetrize these functions in the coordinetes of the electrons (holes) keeping the correct normalization. (The particle coordinates are not explicitly apearing in our short-hand notations.) With the help of the basis (4.28) one reduces the solution of the stationary two-electron-hole-pair Schrödinger equation to a matrix diagonalization. The detailed form of the matrix elements is given in Hu (1991).

Using the one- and two-pair matrix diagonalization results for the respective ground-state energies it is possible to compute the molecular binding energy, Eq. (3.25). Examples of the results are plotted in Fig. 4.6 for three different electron-hole mass ratios and $\epsilon_2/\epsilon_1 = 1$. The full lines are obtained using the numerical matrix diagonalization technique and the dashed curves are computed using third-order perturbation theory. The figure shows, that independent of the electron-hole mass ratio, the molecular binding energy increases with decreasing dot radius. For the physically unrealistic but theoretically interesting limit $R \rightarrow 0$, we see that the ground-state biexciton energy approaches values of one to two times the exciton binding energy in the bulk. The molecular binding energy remains always positive and increases for small radii. Furthermore, Fig. 4.6 clearly shows the good asymptotic agreement with the Coulomb perturbation theory results. The dependence of the molecular binding energy on the ratio of background dielectric constants ϵ is illustrated in Fig. 4.7 (for

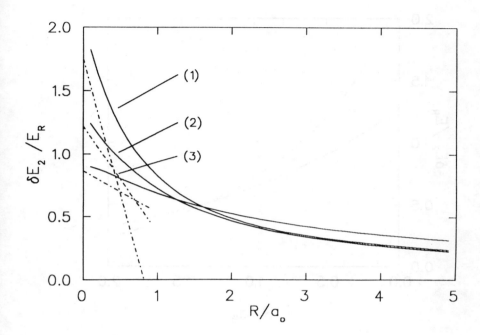

Fig. 4.6: The biexciton binding energy for a mass ratio m_e/m_h = 0.1 (curve 1), 0.2 (curve 2) and 1 (curve 3). The broken lines show the results of third-order perturbation theory. From Hu *et al.* (1990b).

m_e/m_h = 0.174 and infinite confinement barriers).

The computed wavefunctions describe an important modification of the electron and hole probability distributions as a consequence of the electron-hole asymmetry which arises through quantum confinement and Coulomb interaction. This effects is best illustrated by computing the radial distribution of electrons and holes in the exciton, defined (for an $L = 0$ state) as

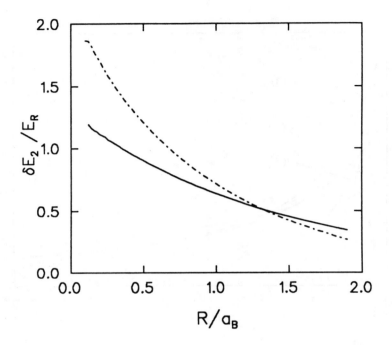

Fig. 4.7: The biexciton binding energy for two different relative dielectric constants: $\epsilon = \epsilon_1/\epsilon_2 = 1$ (full line) und $\epsilon = 10$ (broken line). From Hu *et al.* (1990c).

$$P_{e,h}(r_{e,h}) = r_{e,h}^2 \int d^3x_{h,e} \, |\psi(\mathbf{x}_e, \mathbf{x}_h)|^2 \, . \tag{4.30}$$

Fig. 4.8 compares examples of the results for a mass ratio of $m_e/m_h = 0.24$ computed in the presence of Coulomb interactions with radial distribution functions obtained for the ideal strong confinement limit, i.e., without Coulomb interaction. The essential new fact is the separation of the local charge distribution as a consequence of the Coulomb correlations. This charge separation yields a net local charge density. The heavier holes are pushed toward the center, similar to the features assumed in the treatment of the intermediate confinement configuration. Since many of the semi-

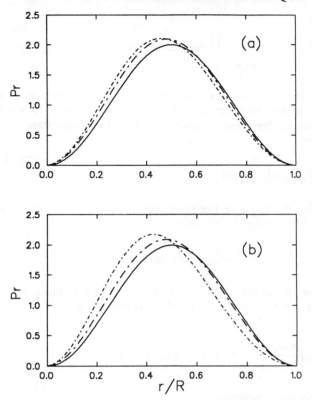

Fig. 4.8: Radial distribution of the hole (short dashed) and of electrons (long dashed line) for R/a_B = 0.5 (a) and R/a_B = 1.0 (b). The full line shows the distribution in the absence of the Coulomb interaction. From Hu *et al.* (1990b).

conductor materials used for quantum dots are highly polar, a local charge density leads to important interaction phenomena with the ionic lattice (LO phonons), which will be discussed in Chap. 7.

As another result of these numerical calculations one may compute the oscillator strengths of optical transitions. Since the carrier-carrier Coulomb interaction depends on the separation between the carriers, the original spherical symmetry of the non-interacting electrons and holes in a quantum dots is broken. This symmetry reduction makes transitions dipole allowed, which would be forbidden in the case without Coulomb interaction.

To illustrate this point let us consider the creation of a biexciton from the lowest exciton state, characterized by the polarization matrix element

$$\langle 1_0 | \, P \, | 2 \rangle = p_{cv} \sum_{n\ell m} \langle 1 | \, a_{(e)n\ell m} \, a_{(h)n\ell - m} \, | 1 \rangle \; . \qquad (4.31)$$

In the strong confinement limit the one- and two-electron-hole-pair states $| 1_0 \rangle$, $| 2 \rangle$ are in the Fock space obtained through the application of products of the respective creation operators on the vacuum state. The exciton ground state is

$$| 1_0 \rangle = a^\dagger_{(e)100} \, a^\dagger_{(h)100} \, | 0 \rangle \qquad (4.32)$$

and, due to the structure of the interband polarization operator, Eq. (4.31), the only possible biexciton state is

$$| 2 \rangle = a^\dagger_{(e)n\ell m} \, a^\dagger_{(h)n\ell - m} \, a^\dagger_{(e)100} \, a^\dagger_{(h)100} \, | 0 \rangle \; . \qquad (4.33)$$

The successive transition energies are then the same as those for creation of a single pair from the vacuum. The lowest energy is

$$E_g + E_R \left(\frac{\pi \, a_B}{R} \right)^2 \quad (n = 1, \; \ell = 0, \; m = 0) \qquad (4.34)$$

and the next higher energy is

$$E_g + E_R \left(\frac{4.474 \, a_B}{R} \right)^2 \quad (n = 1, \; \ell = 1, \; m = 0, \pm 1) \; . \qquad (4.35)$$

On the other hand, one can build up another two pair state also from two $\ell = 0$ electrons and two $\ell = 1$ holes. The energy of this state is

$$E_g + E_R \left(\frac{a_B}{R} \right)^2 \left[\frac{\mu}{m_e} \pi^2 + \frac{\mu}{m_h} (4.474)^2 \right] . \qquad (4.36)$$

This state cannot be reached by applying the transition matrix element (4.32) because this forces electrons and holes to have pairwise the same ℓ quantum number. Because only the heavier holes are in a higher state, $\ell = 1$, the energy of this state is significantly lower than that of the second dipole allowed state. In the presence of Coulomb interactions the $L = 0$

biexciton state that correspond to this unperturbed state has an energy which differs from the above expression only through terms of order a_B/R. However, it may have components of the allowed type and therefore causes the appearance of a "forbidden" transition close to the first allowed one. Numerical examples of such states are discussed in Chap. 5 in the context of the nonlinear optical properties of semiconductor quantum dots.

Generally, the matrix diagonalization method outlined in this section is especially reliable for not too big quantum dots since large radii require substantially increased numbers of basis functions, reaching the limits of the computer. Due to the large number of close levels in such large quantum dots it is more advantageous to formulate the problem in a different form, see Sec. 3-3 and Chap. 6.

4-3. Quantum Monte Carlo Technique

Another numerical method which has been applied to compute exciton and biexciton ground-state properties in quantum dots [Pollock and Koch (1991)] is the "quantum Monte Carlo" (QMC) method. Its central object is the many-body density matrix at a given temperature, $k_B T = 1/\beta$,

$$\rho(\mathbf{r}, \mathbf{r}'; \beta) = \sum_n e^{-\beta E_n} \psi_n^*(\mathbf{r}) \psi_n(\mathbf{r}') = \langle \mathbf{r} | \exp(-\beta \mathscr{H}) | \mathbf{r}' \rangle \ . \tag{4.37}$$

The notations here are simplified by denoting all the coordinates of the particles by a single symbol $\mathbf{r}$ and the ensemble of quantum numbers by n. Generally, QMC calculations yield system properties at a finite temperature and ground-state properties are obtained in the large β (low temperature) limit [Pollock and Ceperley (1984), Pollock (1988)].

The numerical QMC algorithm is based on the identity

$$\langle \mathbf{r} | \exp(-\beta \mathscr{H}) | \mathbf{r}' \rangle = \langle \mathbf{r} | \ [\exp(-\tau \mathscr{H})]^L \ | \mathbf{r}' \rangle$$

$$= \int d^3 r_1 \ ... \ d^3 r_{L-1} \ \langle \mathbf{r} | \exp(-\tau \mathscr{H}) | \mathbf{r}_1 \rangle \ ... \ \langle \mathbf{r}_{L-1} | \ \exp(-\tau \mathscr{H}) | \mathbf{r}' \rangle \ . \tag{4.38a}$$

In terms of the density matrix this relation can be written as

$\langle \mathbf{r}|\exp(-\beta \mathcal{H})|\mathbf{r}'\rangle$

$$= \int d^3r_1 \dots d^3r_{L-1}\ \rho(\mathbf{r}, \mathbf{r}_1; \tau) \dots \rho(\mathbf{r}_{L-1}, \mathbf{r}; \tau) . \qquad (4.38b)$$

Here L is an integer and $\tau = \beta/L$ corresponds to a temperature which is L times higher than the original temperature. The "path" in Eq. (4.38) is the sequence of "points" $\mathbf{r}, \mathbf{r}_1, \dots, \mathbf{r}_{L-1}$. Eq. (4.38) is an exact identity, but it is also very useful for an approximate analysis. For a sufficiently small τ one may make a high temperature approximation for the density matrix $\rho(\mathbf{r}, \mathbf{r}'; \tau)$, insert the result into Eq. (4.38) and evaluate the multidimensional integral. For the integration one usually uses the Monte Carlo integration method. This is the reason why this procedure is called "Quantum Monte Carlo method".

Besides evaluating the multidimensional integration, the basic work in QMC studies is a good choice of the high temperature density matrix. The simplest approximation is based on the Trotter formula [Pollock and Runge (1992)]

$$e^{-\tau(T + W)} = e^{-\tau W/2}\ e^{-\tau T}\ e^{-\tau W/2} + \mathcal{O}(\tau^3) , \qquad (4.39)$$

where T symbolizes the kinetic part of the Hamiltonian and W is the interaction part, respectively, so that

$$\mathcal{H} = T + W . \qquad (4.40)$$

For the density matrix Eq. (4.39) translates to

$$\rho(\mathbf{r}, \mathbf{r}'; \tau) = \langle \mathbf{r}|\exp(-\tau \mathcal{H})|\mathbf{r}'\rangle = \rho^{(1)}(\mathbf{r}, \mathbf{r}'; \tau)\ e^{-\tau[W(\mathbf{r})+W(\mathbf{r}')]/2} + \mathcal{O}(\tau^3) , \qquad (4.41)$$

where

$$\rho^{(1)}(\mathbf{r}, \mathbf{r}'; \tau) = \langle \mathbf{r}|e^{-\tau T}|\mathbf{r}'\rangle \qquad (4.42)$$

is the free particle part of the density matrix. It can be computed from the equation

$$\frac{\partial}{\partial \tau} \rho^{(1)}(\mathbf{r}, \mathbf{r}'; \tau) = -T\ \rho^{(1)}(\mathbf{r}, \mathbf{r}'; \tau) = \frac{\hbar^2 \nabla^2}{2m}\ \rho^{(1)}(\mathbf{r}, \mathbf{r}'; \tau) . \qquad (4.43)$$

For perfect quantum confinement Eq. (4.43) has to be solved with the

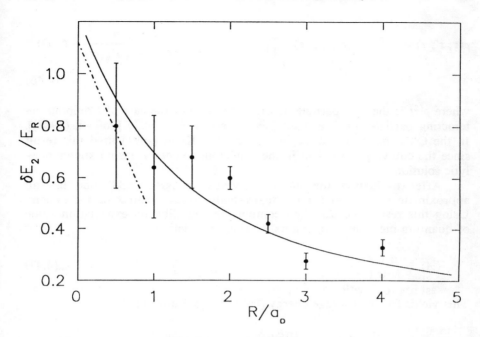

Fig. 4.9: Calculated binding energy of the biexciton for $m_e/m_h = 0.24$, $\epsilon = 1$ and infinite confinement potential. Results of the matrix-diagonalization (full line), of the Coulomb perturbation theory (broken line), and of the Quantum-Monte-Carlo method (dots with error bars) are shown [Hu *et al.* (1990a), Pollock and Koch (1991)].

boundary condition

$$\rho^{(1)}(\mathbf{r}, \mathbf{r}'; \tau) = 0 \ \text{ for } r, r' \geq 0 \tag{4.44}$$

and the "initial" condition

$$\rho^{(1)}(\mathbf{r}, \mathbf{r}'; 0) = 0 \ . \tag{4.45}$$

As discussed by Pollock and Koch (1991), Eq. (4.43) can be solved analytically, but since this solution does not give additional insights for the present discussion we suppress it here and refer the interested reader to the original literature.

Instead of using Eq. (4.41), in QMC calculations one often makes a high temperature cluster approximation

$$\rho(\mathbf{r}, \mathbf{r'}; \tau) = \prod_i \rho^{(1)}(\mathbf{r}_i, \mathbf{r'}_i; \tau) \prod_{j<k} \frac{\rho^{(2)}(\mathbf{r}_j, \mathbf{r}_k, \mathbf{r'}_j \mathbf{r'}_k; \tau)}{\rho^{(1)}(\mathbf{r}_j, \mathbf{r'}_j; \tau)\rho^{(1)}(\mathbf{r}_k, \mathbf{r'}_k; \tau)} + \mathcal{O}(\tau^3),$$

$$(4.46)$$

where $\rho^{(2)}$ is the two-particle density matrix. For the case of Coulomb interacting particles in free space $\rho^{(2)}$ is known analyticaly [Pollock (1988)]. In the QMC quantum dot study Pollock and Koch (1991) used this result since the equivalent problem in the confining geometry has no known analytic solutions.

After evaluating the multidimensional integral (4.38) one has an approximate expression for the many-body density matrix of the system. Using this result one can then compute the equilibrium expectation value of quantum mechanical operators A using the relation

$$\langle A \rangle = \frac{\text{tr}(A \rho)}{\text{tr}(\rho)}.$$

$$(4.47)$$

This yields for the average energy [Pollock and Runge (1992)]

$$E = \langle \mathscr{H} \rangle = \frac{\text{tr}(\mathscr{H} \rho)}{\text{tr}(\rho)} = -\frac{\text{tr}(\partial\rho/\partial\beta)}{\text{tr}(\rho)} = -\frac{\text{tr}(\rho \, \partial\ln\rho/\partial\beta)}{\text{tr}(\rho)} = -\langle \partial\ln\rho/\partial\beta \rangle.$$

$$(4.48)$$

Depending on the Hamiltonian, this expression yields the exciton energy or the biexciton energy, respectively.

QMC results for the exciton and biexciton ground-state energies have been obtained [Pollock and Koch (1991)] and used to compute the molecular binding energy. The results are compared in Fig. 4.9 to the matrix-diagonalization results and to perturbative calculations. The quoted uncertainty in the Monte Carlo results is a combination of the statistical uncertainty in the Monte Carlo calculation and the uncertainty in extrapolating to the small τ, large β limits. Within the errors, the overall agreement of the different mutually independent calculations is quite good. In order to use the QMC result for the density matrix, one has to evaluate the trace in real space representation. Inserting for A the density operator allows to compute the electron and hole density profiles (radial distributions) in the quantum dot. Examples of the results are shown in Figs. 4.10 and 4.11, respectively.

In agreement with the matrix-diagonalization results discussed in the previous section, also the QMC calculations show a significant difference between the electron and hole distributions, which becomes more pronounced for smaller quantum dots, compare Figs. 4.10a, 4.11a, and 4.11c. In all cases we observe significant deviations from the noninteracting parti-

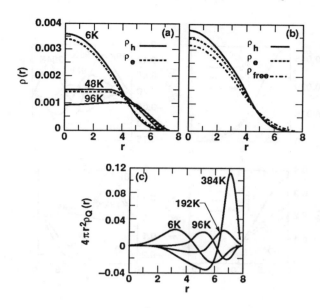

Fig. 4.10: Density profiles of a single electron-hole pair in a quantum dot with $R = 8\ a_B$. (a) The electron (dashed line) and hole (solid line) profile is plotted for various temperatures. (b) The density profiles for an interacting electron-hole pair at $T = 6$ K are compared to the noninteracting result (ρ_{free}). (c) The charge densities $\propto (\rho_h - \rho_e)$ are plotted corresponding to the density profiles in (a). From Pollock and Koch (1991).

cle results.

The significance of the local charge separation, which is given by the differences between the local electron and hole distribution, was already pointed out in the previous section. As will be discussed in Chap. 7, the local charge density leads to an enhanced coupling of electron-hole pairs to LO phonons. The QMC calculations allow to investigate the equilibrium charge separation as function of the system temperature. The examples plotted in Figs. 4.10c, 4.11b, and 4.11d show a very interesting trend. At low temperatures, where only the ground state is occupied, the lighter electrons cushion the holes away from the surface of the dot, leading to an accumulation of positive charge around the dot center and of negative charge close to the surface of the dot. With increasing temperature, electrons and holes are no longer in a pure ground-state distribution since also the excited states are partially (thermally) populated. This effect is more

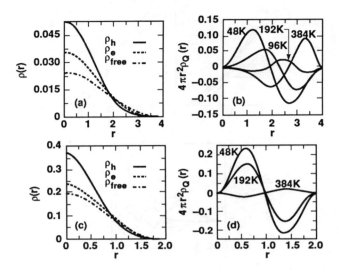

Fig. 4.11 Density profiles of a single electron–hole pair in a quantum dot with $R = 4\, a_B$ (a) and $R = 2\, a_B$ (c) in comparison to the noninteracting result (ρ_{free}). The charge densities $\propto (\rho_h - \rho_e)$ for $R = 4\, a_B$ and $2\, a_B$ are shown in (b) and (d), respectively. From Pollock and Koch (1991).

prominent in larger quantum dots, since the separation between the quantum confined energy levels is smaller. Fig. 4.10c shows that with increasing temperature a charge inversion occurs in dots with $R = 8\, a_B$. The holes move closer to the dot surface pushing the electrons closer to the center. This trend is also present in smaller dots, Figs. 4.11b and 4.11d, but the charge inversion occurs at higher temperatures due to the reduced thermal population of the excited states.

4-4. Surface Polarization Instabilities

So far we discussed mostly quantum dots with an infinite potential barrier. For the few cases where we considered finite potential barriers we ignored the mismatch between the dielectric constants inside and outside the dots, assuming that the ratio (2.61) of the dielectric constants $\epsilon = 1$. In this section we now extend our discussion for finite potential wells by including the dielectric mismatch. We will see that the combination of the two effects leads in some paramter range to substantial modifications of the energy structure and charge distribution in the quantum dots.

The classical treatment of the dielectric surface polarization in Sec. 2-5 shows that selfenergy contributions arise which are singular on the surface of the dot. With an infinite confinement barrier the Schrödinger equation nevertheless remains well-defined since the singularity is compensated by the vanishing of the wavefunction. However, for the more realistic case of a finite potential well the singularity is not integrable and allows no normalizable ground-state solution of the Schrödinger equation. The particle wavefunctions collapse to the outer side of the surface. These unphysical results clearly show that the classical electrostatic description of the interface between dielectric media fails at distances comparable to the interatomic distance.

To avoid an extremely complicated microscopic treatment of the surface dielectric effects, it is compelling to chose a similar heuristic approach as that taken for the analogous problem which occurs in semiconductor quantum wells [Tran Thoai *et al.* (1990b)]. Here, one introduces a phenomenological cut-off distance of the same order of magnitude as the interatomic distance, which regularizes the potential. With such a cut-off Tran Thoi *et al.* (1990b) could show for GaAs-GaAlAs quantum wells that, even though the confinement potential barriers are low, the corrections due to the dielectric surface polarization are nevertheless small in this system, as a consequence of the small difference of the dielectric constants.

However, for quantum dots embedded in glass or in a liquid the dielectric constants inside and outside the dot are very different, and therefore the surface polarization effects are expected to be important. To illustrate these features, we discuss here the ground state of the electron-hole pair in a quantum dot with finite potential barrier and surface dielectric effects [Bányai *et al.* (1992)]. To simplify the discussion we assume the same effective masses inside and outside the dot. For convenience, we introduce again the normalized coordinates $x \equiv r/R$.

The general term of the series defining the selfenergy (see Sec. 2-5)

$$\Sigma(x) = \frac{1}{2} \, \delta V(\mathbf{x}, \mathbf{x}) \tag{4.49}$$

behaves for $\ell \to \infty$ like the geometrical series, showing that the selfenergy is diverging as $x \to 1$. Subtracting known series, one may put the selfenergy into the form

$$\Sigma(x) = E_R \frac{a_B}{R} \frac{\epsilon - 1}{\epsilon + 1} \left[\frac{1}{1 - x^2} - \frac{1}{x^2} \frac{\epsilon}{\epsilon + 1} \ln(1 - x^2) \right.$$

$$\left. + \frac{\epsilon^2}{\epsilon + 1} \sum_{\ell=0}^{\infty} \frac{x^{2\ell}}{(\ell + 1)[1 + \ell (\epsilon + 1)]} \right] \tag{4.50a}$$

for $x < 1$ and

$$\Sigma(x) = E_R \frac{a_B}{R} \frac{\epsilon-1}{\epsilon+1} \epsilon \left[\frac{1}{1 - x^2} - \frac{1}{\epsilon+1} \ln(1 - x^{-2}) \right.$$

$$\left. + \frac{1}{x^2} \frac{\epsilon}{\epsilon+1} \sum_{\ell=0}^{\infty} \frac{x^{-2\ell}}{(\ell+1)[1 + \ell (\epsilon+1)]} \right] \tag{4.50b}$$

for $x > 1$. After this separation of the singularities, the remaining series are converging everywhere, including the boundaries. In Fig. 4.12 we show the selfenergy in a radial domain ranging from the center of the sphere to twice its radius.

The divergence at the surface of the sphere is not integrable and is pathological for the Schrödinger equation. The attractive potential just outside the sphere captures (traps) all particles at $r = R + 0$. In what follows we therefore use a regularized selfenergy $\Sigma^r(x)$, which is finite everywhere. It coincides with $\Sigma(x)$ for $|x - 1| > \delta \, a_B/R$ and in the interval $|x - 1| < \delta \, a_B/R$ the regularized selfenergy is just a linear interpolation between its values at $x = 1 \pm \delta \, a_B/R$, respectively. The cut-off parameter δ represents the exclusion of a layer whose thickness should be of the order of the lattice constant. Electrodynamics of continuous media breaks down in this domain and, in principle, one should use microscopic theory. The series defining the interaction energy are also diverging for $r_1 \rightarrow R$ and simultaneously for $r_2 \rightarrow R$, however, these divergencies are integrable and thus not dangerous for the Schrödinger equation.

Besides their Coulomb interactions, the electron and hole in a quantum dot feel a confinement potential, chosen here as a finite potential barrier of height U_0. We are interested to find the lowest-lying energy eigenvalues and wavefunctions of the Hamiltonian

$$\mathcal{H} = -\frac{\hbar^2}{2m_e} \nabla_e^2 - \frac{\hbar^2}{2m_h} \nabla_h^2 + U_0 [\, \theta(r_e - R) + \theta(r_h - R) \,] + W^r(\mathbf{r}_e, \mathbf{r}_h) \tag{4.51}$$

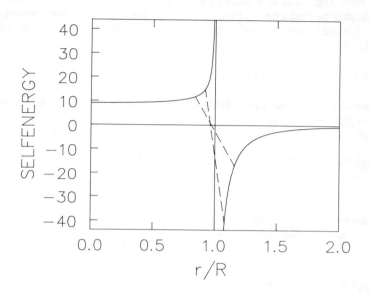

Fig. 4.12 Selfenergy of a charged particle (in units of the Rydberg energy E_R at the radial position x (in units of the sphere radius R), inside and outside the dielectric sphere having a relative dielectric constant to the surrounding $\epsilon = 10$. The dashed lines correspond to regularizations with $\delta = 0.08$ and 0.16.

for an electron-hole pair with effective electron and hole masses m_e and m_h, respectively. The superscript "r" at the effective Coulomb energy indicates that the selfenergy of Sec. 2-5 was replaced by the regularized one. Actually, for the numerical evaluations it is more convenient to work with a continous potential barrier which is linearly interpolated between 0 an U_0 in the interval $R - \delta\, a_B < r < R + \delta\, a_B$ and not with the usual step function barrier $U_0\, \theta(r-R)$.

On general grounds one may expect that if the depth of the dielectric Coulomb trap just outside the dot is much larger than the height of the confinement barrier, then the particles may be trapped in this potential minimum. The first one to be trapped would be the heavier particle, i.e., the hole. To investigate this scenario and to find the range of physical parameters where this surface trapping occurs, Bányai *et al.* (1992) solved the Schrödinger equation with the two-particle Hamiltonian (4.51) using two different numerical methods.

The first approach is a Hartree or self-consistent potential approximation. This means that one factorizes the two-body ground-state wavefunction into a simple product of spherically symmetrical one-particle wavefunctions,

$$\Psi(\mathbf{r}_e, \mathbf{r}_h) \simeq \psi_e(\mathbf{r}_e) \, \psi_h(\mathbf{r}_h) \, . \tag{4.52}$$

The variational principle then yields the self-consistent coupled one-particle Schrödinger equations

$$\left[- \frac{\mu}{m_e} a_B^2 \, E_R \, \nabla_e^2 + U_0 \, \theta(r_e - R) + \Sigma^r(r_e/R) - U_e(r_e) - E_e \right] \psi_e(r_e) = 0 \tag{4.53}$$

for the electron and analogously for the hole. The potential energy of the
electron in the average field of the hole is given by

$$U_e(r_e) = \int d^3r_h \, W(\mathbf{r}_h/R, \mathbf{r}_e/R) \, |\psi_h(r_h)|^2 \, . \tag{4.54}$$

Due to the s-symmetry of the one-particle ground state functions only the
$\ell = 0$ partial wave of the interaction potential contributes here. The total
ground state energy E differs from the sum of the one particle Hartree
ground state energies through the Hartree constant,

$$E = E_e + E_h + \int d^3r_e \, d^3r_h \, W(\mathbf{r}_e/R, \mathbf{r}_h/R) \, |\psi_e(r_e)|^2 \, |\psi_h(r_h)|^2 \, . \tag{4.55}$$

The system of one-dimensional Schrödinger equations was solved iter-
atively for the two particles using the Runge-Kutta bisection method for
one-dimensional Schrödinger equations. The iteration starts by solving the
Schrödinger equation of the lighter particle (electron) without interaction
with the heavier particle (hole). This wavefunction is used in the full
equation of the heavier particle and so on. It turns out that convergence is
achieved typically within five steps.
     An independent check of the Hartree approach is obtained by adopt-
ing the matrix diagonalization method of Sec. 4-2 for the problem of finite
boundaries. The main modification of this method, when applied to the
problem under discussion, is the choice of the radial part of the one-parti-

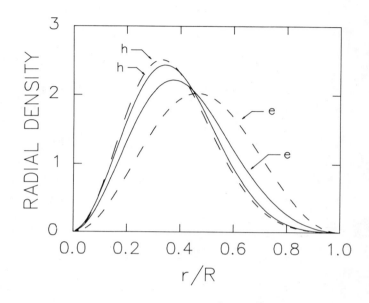

Fig. 4.13: Radial distribution of electrons and holes in an infinitely high potential well with $R = a_B$, $\epsilon = 10$ and $m_e/m_h = 0.1$ (full line - matrix diagonalization, dashed line - Hartree approximation). From Bányai *et al.* (1992).

cle basis. Since the potential barrier is finite and the ground-state wavefunction is not to be expected to vanish everywhere outside the sphere boundary, Bányai *et al.* (1992) choose the basis of the wavefunctions of a single particle moving freely in a fictive sphere whose radius was twice that of the actual dot. Numerical checks showed that this size of the fictive sphere was large enough to not influence the results.

The one-electron-hole problem was solved numerically and the height of the potential barrier U_0 and the quantum dot radius R were varied in discrete steps in a wide domain to catch the whole scenario of surface trapping. In Fig. 4.13 the radial distribution of electrons and holes for the case of an infinite potential barrier ($U_0 \to \infty$) is shown both for the Hartree approximation and the matrix diagonalization method. Even though one sees some quantitative differences, both methods show that the electron and hole distributions are displaced from another, and the heavier hole is cushioned away from the surface and pushed toward the center of the sphere. For later reference we call a state with such an electron-hole dis-

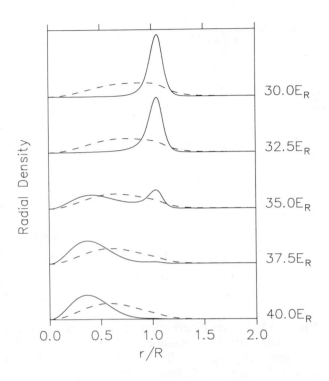

Fig. 4.14: Radial distributions of electrons (dashed curves) and holes (full curves) by different potential barrier heights U_0 (in units of E_R) with $R = a_B$, $\epsilon = 10$, $\delta = 0.08$ and $m_e/m_h = 0.1$, within the Hartree approximation. From Bányai *et al.* (1992).

tribution a "volume state" of the quantum dot.

In Fig. 4.14 the changes of the radial distributions of the electrons and holes introduced by a finite potential barrier are shown within the Hartree approximation. At $U_0 = 40\ E_R$ the barrier acts basically like an infinite one, but a further reduction down to $30\ E_R$ causes gradual changes of the hole position. The hole gradually moves to the surface of the sphere, while the electron still remains delocalized inside the sphere. A further decrease of the potential barrier brings the electron also to the surface, but still with a wider distribution than that of the hole. We denote such as state as "surface state". Qualitatively this picture survives even after the complete elimination of the confinement barrier ($U_0 = 0$). The particles remain con-

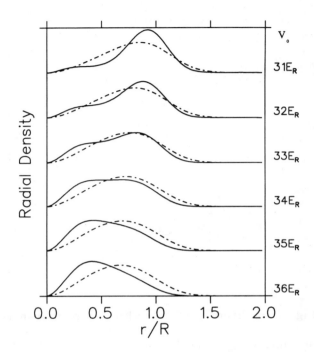

Fig. 4.15: Radial distributions of electrons (dashed curves) and holes (full curves) by different potential barrier heights U_0 (in units of E_R) with $R = a_B$, $\epsilon = 10$, $\delta = 0.08$ and $m_e/m_h = 0.1$, within the matrix diagonalization method. From Bányai *et al.* (1992).

fined in the minimum of the potential energy given by the selfenergy represented in Fig. 4.12.

Essentially the same scenario is described by the matrix diagonalization calculations, but Fig. 4.15 shows that some quantitative differences occur. The hole radial distribution on the surface is less sharp and the electron radial distribution follows more closely that of the holes. As in the Hartree approximation one observes a gradual transition of the electron-hole pair from volume to surface state.

In Fig. 4.16 the energies of the electron-hole pair ground state as function of confinement potential height are shown. It can be seen, that parallel to the localization of hole and electron on the surface, a strong

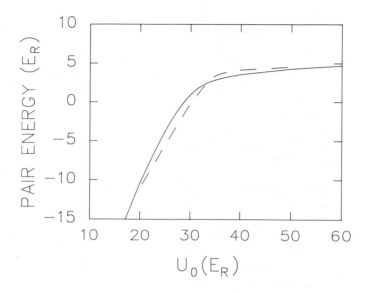

Fig. 4.16: Energies of the calculated electron-hole pair states for different potential barrier heights U_0 (in units of E_R) with $R = a_B$, $\epsilon = 10$, $\delta = 0.08$ and $m_e/m_h = 0.1$, within the matrix diagonalization method (full curve) and the Hartree approximation (dashed curve). From Bányai *et al.* (1992).

drop of the energy follows, which soon becomes negative. Negative energies here mean a reduction of the electron-hole-pair ground state in the quantum dot below the bandgap of the corresponding bulk semiconductor material. This energy reduction would amount to a red shifted absorption onset in the quantum dot in comparison to the continuum absorption in bulk. The polarization energy compensates and even overcompensates the confinement energy. Before localization of both particles on the surface sets in, the matrix diagonalization results are lower than the Hartree results. For smaller confinement potential values the Hartree calculations give lower energy values. In principle the matrix diagonalization energies can always be reduced to less or equal the Hartree calculation values simply by increasing the expansion basis. In the present case this numerically increasingly expensive task was not attempted, since there is no reason to believe that the domain of negative energies has physical significance.

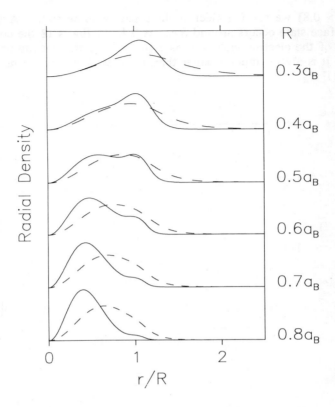

Fig. 4.17: Radial distributions of electrons (dashed curves) and holes (full curves) as functions of the normalized radius r/R for dots of different radii R for the same parameters $U_0 = 40\ E_R$, $\epsilon = 10$, $\delta = 0.08$ and $m_{e/m_h} = 0.1$ within the Hartree approximation. From Bányai *et al.* (1992).

In an experimentally realized quantum dot system it is not trivial to change the confinement potential, since this is basically determined by the chosen semiconductor and host materials. However, a reduction of the dot size for fixed confinement potential also leads to a decreasing influence of the quantum confinement and to a more pronounced penetration of the hole and electron into the host material. To investigate this experimentally relevant situation we show in Fig. 4.17 computed electron and hole distribution functions (within the Hartree approximation) for various dot sizes at fixed potential barrier ($U_0 = 40\ E_R$ and $\delta = 0.08$). For large dot radii

$(R/a_B \geq 0.8)$ we see the electron-hole pair volume state. A transition to the surface state occurs around $R/a_B \simeq 0.4$. In Fig. 4.18 the corresponding energy of the electron-hole pair as a function of the dot radius R is represented. It is always much smaller than the ideal kinetic confinement energy $(\pi a_B/R)^2 \, E_R$.

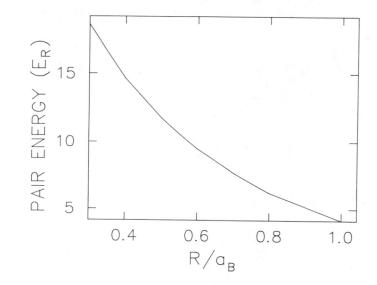

Fig. 4.18: Energies of the calculated electron–hole pair states for different dot radii R with $U_0 = 40 \, E_R$, $\epsilon = 10$, $\delta = 0.08$ and $m_e^*/m_h = 0.1$ within the Hartree approximation. From Bányai *et al.* (1992).

These numerical results show that the interplay between the attractive surface polarization and the repulsive quantum confinement potential leads to a scenario where the electron-hole excitations may be confined inside the quantum dot or trapped close to the surface. For typical CdS parameters ($a_B = 30 \, \overset{\circ}{A}$ and $E_R = 30 \, \text{meV}$) and regularization of the surface potential at a reasonable cut-off distance (about $5 \, \overset{\circ}{A}$), surface trapping of the optically generated electron-hole pair occurs at a confinement potential height of the order of $30 \, E_R$. Alternatively, for fixed confinement potential ($40 \, E_R$) the transition from the volume to surface state occurs around $R/a_B \simeq 0.4$.

The scenario described is a direct consequence of the use of the continuum theory for the dielectric boundary effects. The resulting divergence of the particle selfenergies is mainly a difficulty of the theory itself. The simplification of using the same effective masses outside the sphere may influence quantitatively the results, however the strong attractive potential just outside the dots remains. Even though the actual numbers resulting from this simplified study may have to be considered with a grain of salt, the qualitative aspects of the theory should be relevant for the understanding of experimental observations. Here we think in terms of a scenario with a possible dielectric self-trapping in small quantum dots, mainly of the hole, close to the surface of the dot, whereas the electron is still mainly delocalized inside the dot. In this way the confinement energy roughly corresponds to the elementary kinetic confinement energy, while the charge distributions of the two particles are significantly separated and the probability for the hole to tunnel out of the dot into the surrounding material is increased.

4-5. SPIN-ORBIT COUPLING

An often followed procedure in solid state physics is to first drastically simplify the model of the system under consideration and then later to add more and more complicated improvements. Usually it is not practical to implement more than a few sophisticated features at once, and one has to study the influence of different effects seperately. In this section we follow this general procedure and discuss modifications of the quantized electron and hole states which occur as a consequence of band mixing effects caused by the quantum confinement.

In the previous sections we treated the spin of the electrons and holes as an irrelevant degeneracy index, i.e., the spin part of the wavefunctions was supposed to be independent of the orbital (real space) motion. Unfortunately, this simple picture is known to be incorrect in most of the compound semiconductors. Already simple $k \cdot p$ perturbation theory [Luttinger (1956), Kane (1957)] shows that the relativistic spin-orbit interaction leads to a mixing of the spin states with the orbital states of the valence bands. This mixing leads to significant energy splittings of the six, otherwise at k = 0 degenerate valence bands. Recently it was shown [Xia (1989)] that these effects may also be important for the description of semiconductor quantum dots.

We discuss in the following the general situation where the valence band wavefunctions in the absence of the spin-orbit mixing are six-fold degenerate. This six-fold degeneracy results from the three-fold degeneracy, corresponding to the p-type behavior of the Bloch part of the wavefunction at k = 0 (Γ-point of the Brillouin zone), and the two-fold degeneracy due to the spin degree of freedom. Since the conduction band is s-type, it has only the two-fold spin degeneracy.

At the Γ-point the spin-orbit interaction conserves only the total angular momentum composed of the orbital and spin parts. Therefore, the spin-orbit mixed states have to be classified according to the irreducible representations of the group of simultaneous rotations of coordinates and spins. From orbital angular momentum 1 and spin angular momentum 1/2 one may form two degenerate states of total angular momentum $J = 1/2$ and four degenerate states of total angular momentum $J = 3/2$. These two groups of states usually exhibit a significant energy splitting at the Γ-point. In GaAs and ZnSe for example the $J = 3/2$ bands are higher than the $J=1/2$ bands, whereas in CuCl the inverse case is realized. For $k \neq 0$ further splitting occurs, not only between the $J = 1/2$ and $J = 3/2$ bands, but also within the $J_z = \pm 1/2$ and $J_z = \pm 3/2$ subbands. In CdS, due to the presence of crystal field terms, whose symmetry is only cubic and not spherical as assumed in the above described idealization, complete band splitting occurs at the Γ-point. Due to this complication, which does not allow a description within an isotropic model, this case will receive only occasional comments in our following discussion. See for example the book of Bastard (1988) for a detailed discussion of valence-band-structure details relevant to confinement problems.

Depending on the semiconductor material one must adhere to one or another simplifying approximation. In the case of CuCl it is sufficient to consider the upper lying $J = 1/2$ band. Therefore, the simple model of a single spin-degenerate valence band is appropriate. The case of GaAs or ZnS may be treated by retaining only the upper-lying $J = 3/2$ bands. If not stated otherwise, we are always working within the parabolic effective mass approximation. The effective valence band energy is thus assumed to be quadratic and isotropic in k. It exhibits two distinctly different negative effective masses for the two-fold degenerated $|J_z| =1/2$ (light hole) and $|J_z| = 3/2$ (heavy hole) states,

$$E_v^{(1/2)} = - \frac{\hbar^2 k^2}{2m_0} (1 + \mu) \tag{4.56}$$

and

$$E_v^{(3/2)} = - \frac{\hbar^2 k^2}{2m_0} (1 - \mu) . \tag{4.57}$$

The parameter μ can vary only between 0 and 1 in order to obtain positive effective masses for the holes and to identify the $|J_z| = 3/2$ band with the heavy holes. The energies (4.56) and (4.57) are actually the eigenvalues of the Baldereschi-Lipari spherical Hamiltonian [Baldereschi and Lipari (1973)]

$$\mathcal{H} = -\frac{\hbar^2}{2m_0} \{ \mathbf{p} + \mu [5/4 - (\mathbf{p} \cdot \mathbf{J})^2] \}, \tag{4.58}$$

obtained from the Kohn–Luttinger Hamiltonian [Luttinger (1956)] after omitting the terms that are invariant only with respect to the rotations of the cube. Here, $\mathbf{p} = -i\hbar\nabla$ and $\mathbf{J}$ are the three 4×4 matrices representing the vector components of the 3/2 spin. The matrix Hamiltonian (4.58) is explicitly invariant with respect to the simultaneous rotations of coordinates and spins. Applied to plane waves with wavevector $\mathbf{k}$ it allows easy diagonalization by just choosing the quantization axis of the $\mathbf{J}$–spin along $\mathbf{k}$. Thus, one obtains the two–fold degenerate eigenvalues of Eqs. (4.56) – (4.57).

In the case of a quantum dot with rigid boundaries in the strong confinement limit one has to solve the matrix Schrödinger equation for a hole [Xia (1989)]

$$\mathcal{H}_{\alpha\beta}\, \psi_\beta(\mathbf{r}) = -E_h\, \psi_\alpha(\mathbf{r}) \qquad (\alpha, \beta = 1, 2, 3, 4) \tag{4.59}$$

with the boundary conditions

$$\psi_\alpha(\mathbf{r}) \Big|_{r = R} = 0 . \tag{4.60}$$

Since the boundary conditions conserve the spherical symmetry of the problem, it is to be expected that an adequate use of group theory might simplify this system of equations considerably.

First of all, we note the algebraic identity

$$(\mathbf{p} \cdot \mathbf{J})^2 - \frac{1}{3}\, \mathbf{p}^2 \mathbf{J}^2 = \frac{1}{9} \sum_{i,j} \mathrm{p}_{ij}\, \mathrm{J}_{ij} \tag{4.61}$$

where

$$p_{ij} = 3\, p_i\, p_j - \delta_{ij}\, p^2 \tag{4.62}$$

and

$$J_{ij} = \frac{3}{2} (J_i\, J_j + J_j\, J_i) - \delta_{ij}\, J^2 \tag{4.63}$$

are symmetrical traceless tensors of second rank. Such tensors are irreducible tensors transforming according to the spin-two representation of the SU_2 group, which is isomorphic to the three-dimensional rotation group. There are five independent components which may be chosen to correspond to the eigenfunctions of the z-component of the spin two, labelled

$$M = 0, \pm 1, \pm 2 .$$

The tensor contraction in Eq. (4.61) gives a scalar, i.e. an irreducible tensor transforming according to the spin-zero representation. We may rewrite this scalar in terms of the M-components using the standard angular momentum composition procedure with the help of the Clebsch-Gordan coefficients

$$\sum_{i,j} p_{ij} J_{ij} = \sum_{M} \langle 2,\text{M};2,-M|0,0 \rangle \; P_M^{(2)} J_M^{(2)} . \qquad (4.64)$$

On the other hand, for the M-components one may write down also the Wigner-Eckart theorem for matrix elements of irreducible tensors between states belonging to irreducible representations of the SU_2 group,

$$\langle I, I_z | \; T_M^{(2)} | I', I'_z \rangle = \langle 2,\text{M};I',I'_z | I,I_z \rangle \; (I| \; T^{(2)} \; |I') . \qquad (4.65)$$

This procedure shows that the only independent entities of the theory are the reduced tensor matrix elements

$$(L| \; p^{(2)} \; |L') \quad \text{with} \quad L, L' = 0, 1, 2, \ldots ; \; |L - L'| = 0, 2$$

between states of given orbital angular momentum, since $(3/2| \; J^{(2)} \; |3/2)$ is just a number. The reduced matrix elements of the differential tensor operator $P^{(2)}$ in turn are related to the reduced matrix elements of the vector operator ∇ [Baldareschi and Lipari (1973), Edmonds (1957), Rotenberg et al. (1959)] and therefore given in terms of the radial derivatives $\partial/\partial r$.

The Hamiltonian $\mathcal{H}$ of Eq. (4.58) is diagonal in the eigenstates of the total angular momentum

$$\mathbf{F} = \mathbf{L} + \mathbf{J} . \qquad (4.66)$$

The eigenstates of F, F_z may be constructed by composition of eigenstates of L orbital and $J=3/2$ spin states

$$|F,F_z; L,3/2\rangle = \sum_{M, J_z} \langle L,M;3/2,J_z|F,F_z\rangle \ |L, M\rangle \ |3/2, J_z\rangle \ . \quad (4.67)$$

Since the Clebsch-Gordan coefficients usually allow several independent ways to construct a state F, F_z from orbital angular momentum L states, one has to express the eigenstates of the Hamiltonian in terms of superpositions of these states. We expect that the ground state in the quantum dot has an $L = 0$ component (corresponding to no spin orbit coupling i.e. $\mu = 0$) which implies $F = 3/2$. However, according to Eq. (4.64) and the composition rules of the angular momentum, the same quantum number $F = 3/2$ may be obtained also with the help of $L = 1, 2, 3$ states. Fortunately, however, the reduced matrix elements of $P^{(2)}$, and consequently also the Hamiltonian, imply $\Delta L = 0, \pm 2$ so that the ground state is only a superposition of the two independent states generated from $L = 0$ and $L = 2$,

$$|0\rangle = f^{(0)}(r) \ |3/2,F_z ;0,3/2\rangle + f^{(2)}(r) \ |3/2,F_z ;2,3/2\rangle . \quad (4.68)$$

Xia (1989) denotes these states as $S_{3/2}$. The matrix elements of the Hamiltonian between the two independent components can be expressed through the reduced matrix elements [Baldareschi and Lipari (1973)]

$$\langle 0| \ p^{(2)} \ |0\rangle = 0$$

$$\langle 2| \ p^{(2)} \ |2\rangle = 2\sqrt{\frac{15}{7}} \ \hbar^2 \left(\frac{\partial^2}{\partial r^2} + \frac{2}{r} \frac{\partial}{\partial r} - \frac{6}{r^2} \right)$$

$$\quad (4.69)$$

$$\langle 0| \ p^{(2)} \ |2\rangle = -\sqrt{6} \ \hbar^2 \left(\frac{\partial^2}{\partial r^2} + \frac{5}{r} \frac{\partial}{\partial r} - \frac{3}{r^2} \right)$$

$$\langle 2| \ p^{(2)} \ |0\rangle = -\sqrt{6} \ \hbar^2 \left(\frac{\partial^2}{\partial r^2} - \frac{1}{r} \frac{\partial}{\partial r} \right) .$$

The self-adjointness of these matrix elements is with the scalar product

$$\int_0^1 dr \, \mathrm{r}^2 \, \psi_1^\dagger(r) \, \psi_2(r) \ .$$

Then, after using the tables of Clebsch-Gordan coefficients [Rotenberg *et al.* (1959)] and some algebra, one finds the two coupled Schrödinger equations [Xia (1989)] for the $F = 3/2$ radial wavefunctions $f(r)$ and $g(r)$,

$$\begin{bmatrix} -\dfrac{\hbar^2}{2m_0}\left[\dfrac{\partial^2}{\partial r^2}+\dfrac{2}{r}\dfrac{\partial}{\partial r}\right]-E_h & \mu\,\dfrac{\hbar^2}{2m_0}\left[\dfrac{\partial^2}{\partial r^2}+\dfrac{5}{r}\dfrac{\partial}{\partial r}+\dfrac{3}{r^2}\right] \\[3mm] \mu\,\dfrac{\hbar^2}{2m_0}\left[\dfrac{\partial^2}{\partial r^2}-\dfrac{1}{r}\dfrac{\partial}{\partial r}\right] & -\dfrac{\hbar^2}{2m_0}\left[\dfrac{\partial^2}{\partial r^2}+\dfrac{2}{r}\dfrac{\partial}{\partial r}-\dfrac{6}{r^2}\right]-E_h \end{bmatrix}\begin{pmatrix} f(r) \\ g(r) \end{pmatrix}=0\ .$$

$$(4.70)$$

The lowest energy eigenvalue E_h of this $S_{3/2}$ eigenvalue problem is the ground state energy of the hole.

Fig. 4.19 illustrates the numerical results as function of μ for the energetically lowest three $S_{3/2}$ and $P_{3/2}$ states. The fact that all the eigenvalues vanish for $\mu = 1$ is due to the fact that these states are associated with the heavy hole band whose mass becomes infinite at $\mu = 1$.

It is worthwhile to note, that one would have a proportionality of all the eigenvalues to $1 - \mu$ for a spin diagonal Hamiltonian with the two bands described by Eqs. (4.56) - (4.57). However, with the band mixing Hamiltonian of Eq. (4.59) the situation is essentially different. The ground-state energy is almost constant up to $\mu = 0.7$, which is approximately the experimental value for GaAs and ZnS, and is roughly equal to the lowest confined level of a particle whose inverse mass is the average of the inverse masses of the two bands,

$$E_h \simeq \frac{\hbar^2}{2m_0}\left(\frac{\pi}{R}\right)^2 \ ; \qquad \frac{1}{m_0}=\frac{1}{m_{hh}}+\frac{1}{m_{lh}} \tag{4.71}$$

The real space wavefunction of this hole ground state, however differs in important respects from the confined wavefunctions of a particle without spin-orbit coupling, usch as the conduction-band electrons. First of all, due to the presence of the $L = 2$ component, the wavefunctions have non-trivial angular dependencies and secondly, their radial dependence is different. The angular dependence of the resulting hole distribution is however not an objective defined entity. Indeed, there are $2 \times 3/2 + 1 = 4$ degenerate ground states and any normalized linear combination of them is

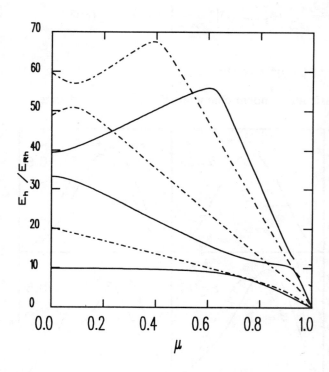

Fig. 4.19 Energy spectrum of $S_{3/2}$ (solid lines) and $P_{3/2}$ (dashed lines) hole states. The energies are measured in units of the hole Rydberg energy. From Hu (1991).

a good ground state as well. The angular dependencies which can be prepared in this way contain three arbitrary continuous parameters corresponding to the continuous three-dimensional rotations. A special wavefunction can be selected only by a special symmetry breaking perturbation. In the absence of such a perturbation we may consider only rotationally invariant entities, which belong to the whole group of degenerate states, independent of the choice of the basis. The radial distribution function of the hole, which can be written as

$$P(r) = r^2\pi \sum_{F_z} |\psi_{3/2,F_z}(\mathbf{r})|^2 = r^2 \int d\Omega \, |\psi_{3/2,F_z}(\mathbf{r})|^2$$

$$= r^2 \left[|f(r)|^2 + |g(r)|^2 \right], \tag{4.72}$$

is such a basis independent characteristic of the ground state.

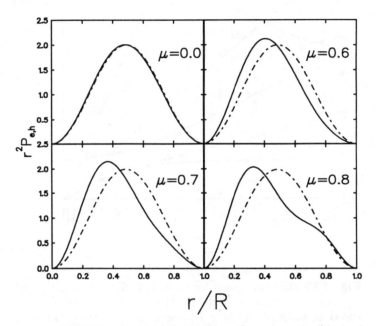

Fig. 4.20: Radial distribution for electron (dashed line) and hole (solid line) for different band coupling constants μ. From Hu (1991).

In Fig. 4.20 we show examples of such radial distributions $P(r)$ for holes and electrons and different μ. With increasing band coupling μ one sees a strong shift of the hole distributions toward the center of the sphere. This radial electron-hole charge separation occurs only through spin coupling and is greater than the charge separation obtained in Sec. 4-2 due to the inclusion of the electron-hole Coulomb interactions. The difference of the radial distribution of the holes and electrons, which is the local radial charge distribution, is represented in Fig. 4.21.

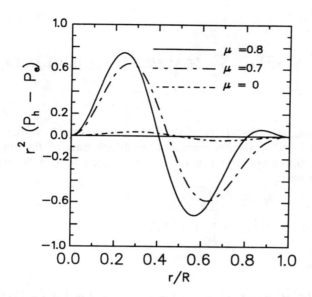

Fig. 4.21: Local radial charge distribution for different band coupling μ. From Hu (1991).

The discussion of optical interband transitions in the presence of the spin-orbit interaction requires some special attention. In the previous chapters we treated a simplified scalar model for the dipole interaction with optical electric fields. In this scalar model it was implicitly assumed that the Bloch part of the wavefunctions takes care of the photon polarization. As soon as we include the spin-orbit coupling, however, we need a more detailed description. For this purpose we assume that the optical field inside the sphere is homogeneous ($\lambda \gg R$) so that the dipole approximation is applicable. The interaction of the system with the field of polarization **e** is then determined by the matrix elements

$$\langle 1/2, \sigma_z, \ell, m | \ \mathbf{d} \cdot \mathbf{e} \ | F, F_z \rangle , \tag{4.73}$$

where **d** is the interband dipole operator. In Eq. (4.73) we characterize the valence band states by F, F_z plus other quantum numbers, and the conduction band state by $1/2, \sigma_z, \ell, m$.

Next, we explicitly separate the radial, angular and Bloch parts of the wavefunctions,

$$| \sigma_z, \ell, m \rangle = f_e(r) \, | \ell, m \rangle \, | 1/2, \sigma_z \rangle \tag{4.74}$$

and

$$|F,F_z\rangle = \sum_L f_h^{(L)}(r) \sum_{M,J_z} \langle L,M;3/2,J_z|F,F_z\rangle \; |L,M\rangle \; |3/2,J_z\rangle \; .$$

(4.75)

Here the summation over L runs through one or two L values. The last kets in (4.74) - (4.75) always represent the Bloch part of the state. Since the interband dipole operator acts only on the Bloch part of the wavefunctions, we may write

$$\langle \sigma_z,\ell,m| \; \mathbf{d} \cdot \mathbf{e} \; |F,F_z\rangle = \int_0^R dr \, r^2 f_e(r) f_h^{(L)}(r)$$

$$\times \sum_{M,J_z} \langle L,M; 3/2,J_z|F,F_z\rangle \; \langle \ell,m|L,M\rangle \; \langle 1/2,\sigma_z| \; \mathbf{d} \cdot \mathbf{e} \; |3/2,J_z\rangle \; .$$

(4.76)

We apply once again the Wigner-Eckart theorem, this time for the vector operator $\mathbf{d}$ (more precisely for its $m = 0, \pm 1$ components) to express its matrix elements through its unique reduced matrix element between the valence and conduction bands :

$$\langle 1/2,\sigma_z| \; d_m \; |3/2,J_z\rangle = \langle 1,m; 3/2,J_z|1/2,\sigma_z\rangle \; \langle 1/2| \, d \, |3/2\rangle \; .$$

(4.77)

With this last step now all matrix elements of the dipole operator are given up to a general interband constant. Due to the mixing of the orbital states of the holes, transitions which are forbidden without the mixing acquire finite oscillator strengths.

Besides other formal presentations [Sweeny and Xu (1989)] of the described spherical model of spin-orbit coupling in quantum dots there is also another different treatment that gives slightly modified results [Sercel and Vahala (1990)]. The starting point of this approach is a spherically symmetrical version of the Kane model [Kane (1957)] that does not decouple the conduction band from the valence band or the split off valence band. This $\mathbf{k}\cdot\mathbf{p}$ model is developed along the lines of a two-particle system consisting of the "envelope" and "Bloch " parts. The two subspaces

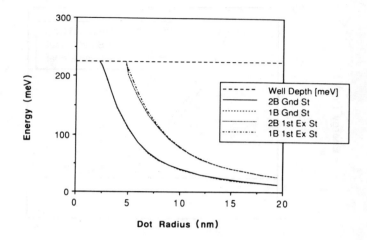

Fig. 4.22: Energies of the lowest conduction-band states for GaAs quantum dots embedded in $Al_{0.3}Ga_{0.7}As$. From Sercel and Vahala (1990).

are coupled through the kinetic energy, since the momentum also has an envelope and a Bloch contribution. The envelope part has to satisfy the confinement conditions imposed by a finite (or infinite) spherical potential barrier.

Numerical calculations within this model are illustrated in Figs. 4.22 - 4.24. In the usual type-I heterostructures (confinement of electrons and holes in the same material) the confinement of the conduction band levels is the same as in the simple one band confinement theory (Fig. 4.22). However, for the case of type-II heterostructures (confinement of electrons and holes in different parts of the heterostructure) there are some deviations (Fig. 4.23). For valence-band levels the situation is more complex as illustrated in the examples shown in Fig. 4.24. Unfortunately, no quantitative comparison of these results and the results based on Xia's (1989) approach is possible on the basis of the published data.

Extensive computations of excitonic, biexcitonic as well as trionic (two holes and one electron) ground states including Coulomb interaction and spin-orbit coupling have been reported in the adiabatic hole motion limit $(m_e/m_h \rightarrow 0)$ using variational approximations [Efros and Rodina (1989)]. Generally, these calculations have the same merits and problems as those without spin-orbit coupling. Especially the computed energy differences have to be treated with care, as discussed in Sec. 4-1.

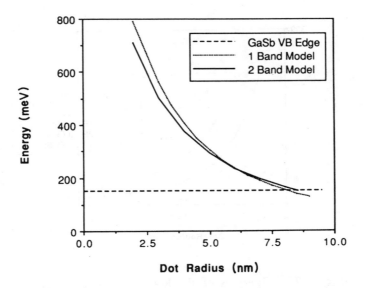

Fig. 4.23 Energies of the lowest lying conduction-band states for InAs quantum dots in GaSb. From Sercel and Vahala (1990).

It is interesting to note, that the inclusion of spin-orbit coupling leads to different wavefunctions even of the noninteracting electron and hole states. Therefore there is a nonvanishing first order perturbation theoretical contribution (see Chapter 3) of the Coulomb interaction to the molecular binding energy. In the limiting case $m_e/m_h \rightarrow 0$ one gets a small negative binding energy of $0.036 \, E_R \, a_B/R$.

4-6. Lattice Model

In all previous chapters the discussion of the quantum dots was performed within the frame of one or another variant of the effective mass theory, where the boundary conditions (confinement) were applied to the envelope of the Bloch functions. This continuum model intuitively seems to be justified as long the linear dimensions of the dots are very large as compared to the crystal lattice constants. In very small microcrystallites, where the number of the "surface" atoms becomes comparable to the number of the "bulk" atoms, this approach becomes very critical.

On the other hand, even within the effective mass theory one my see the necessity for further corrections of the picture in very small crystallites.

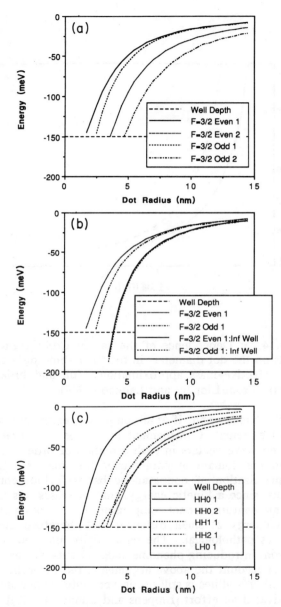

Fig. 4.24 Energies of the highest lying valence-band states for GaAs quantum dots in $Al_{0.3}Ga_{0.7}As$. From Sercel and Vahala (1990).

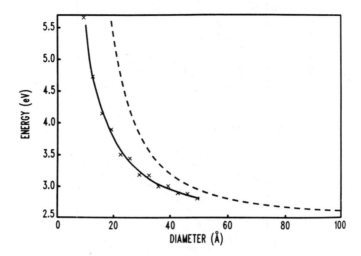

Fig. 4.25: Dependence of the gap of CdS Quantum
Dots on the dot diameter in the "tight-binding"- and
the effective-mass approximations (full and broken
lines). From Lippens and Lannoo (1989).

Indeed, parabolic bands occur, if at all, only in a relatively restricted vicinity of the band extrema. Confinement in a cube of lateral dimension A
implies quantized wave vectors in each direction of order π/A, which may
be well beyond the domain of parabolicity for small A. Unfortunately,
confinement-quantization of nonparabolic bands forbids consideration of
Coulomb effects, since a kinetic energy, that increases less than quadrati-
cally with the momentum, cannot compensate the diverging attractive Coulomb potential energy at small distances and therefore the Schrödinger
equation becomes pathological. The parabolicity assumed so far, actually
has saved us from introducing phenomenological cut-off parameters.

Taking into account the above arguments, the theoretical understanding of very small crystallites is still not in very satisfactory shape. This situation has motivated an effort [Lippens and Lannoo (1989)] to develop an
atomic theory of very small semiconductor crystallites considered as finite
atomic lattices. The eigenvalue problem is formulated as the diagonalization of the one-electron Hamilton operator $\mathscr{H}$ of an N-atom cluster,
written as a matrix in a basis of (hybridized) atomic orbitals,

$$\mathcal{H} = \sum_i \epsilon_i \, a_i^\dagger \, a_i + \sum_{ij} t_{ij} \, a_i^\dagger \, a_j \, . \qquad (4.78)$$

Here, the indices run over all the N ionic positions and over the atomic orbital basis. Taking $N \to \infty$ one obtains a bulk crystal, allowing to fix the parameters ϵ_i and t_{ij} (i,j - nearest neighbors) to fit the known band structure. The N ions are placed on a perfect crystal lattice in a compact cluster. A fit of the band parameters is achieved with the help of thirteen parameters, which implies the inclusion, besides the usual sp^3 basis, also of an s^* excited state. Spin-orbit coupling in the valence band is ignored.

The diagonalization of the Hamiltonian in the bulk limit is easily achieved by the use of discrete Fourier transforms. For a finite lattice, the Fourier transform is of no help and one has to diagonalize an $\nu N \times \nu N$ matrix. The numerical complications of such a problem, where N is 20 to 2500, stems from the need to store all the matrix elements. To avoid this difficulty a recursion method was used [Pettifor and Weaire (1985)].

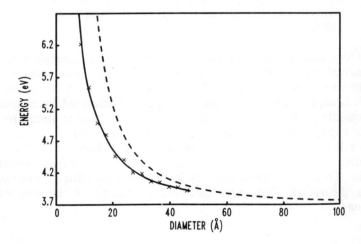

Fig. 4.26 Dependence of the gap of ZnS quantum dots on the dot diameter in the "tight-binding"- and the effective-mass approximations (full and broken lines). From Lippens and Lannoo (1989).

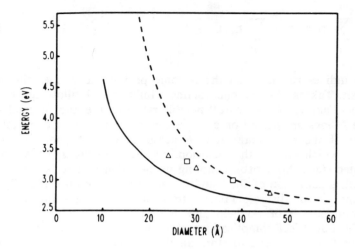

Fig. 4.27 Comparison of the calculated gap of CdS Quantum Dots within the "tight-binding" and effective-mass approximations (full resp. broken lines) with experimental points [Lippens and Lannoo (1989)].

The crystallites are built up by connecting the successive shells of nearest neighbors. Special care is devoted to the crystallite surface since it is well-known, that dangling bonds create localized states in the band gap. In fact it is expected that the dangling bonds are saturated by molecules of the surrounding material. In reality, however, the atomic structure of the surface is experimentally as well as theoretically unknown.

Since it is a pure one-particle theory, no possibility of evaluating the contribution of the electron-hole Coulomb energy exists within this approach. To correct this deficiency in comparisons with experiment, one adds the theoretical first-order Coulomb correction of the electron-hole ground state

$$\delta E_{Coul} = -\ 3.572\ \frac{e^2}{\epsilon d}\ , \tag{4.79}$$

where d is the diameter of the equivalent microsphere. This parameter is defined as

$$d = a \left[\frac{3N}{4\pi} \right]^{1/3} . \tag{4.80}$$

Numerical results for the top of the valence band (TVB) and bottom of the conduction band (BCB) energies have been obtained [Lippens and Lannoo (1989)] for various CdS and ZnS clusters. The tight binding results show oscillations related to the alternative ionic character of the clusters and the limited extension of their interactions. The solid curves in the representations of the data are averaged values.

Figures 4.25 and 4.26 show the computed variations of the band gap for CdS and ZnS, respectively, as function of the cluster diameter. In the same figures the effective mass confinement predictions,

$$E = 2\hbar^2 \frac{\pi^2}{\mu d^2} \; ; \; \left(\frac{1}{\mu} = \frac{1}{m_e} + \frac{1}{m_h} \right) , \tag{4.81}$$

are represented as dashed curves. Due to the unsatisfactory treatment of the valence band, as well as to its anisotropy, the definition of the heavy hole valence band effective mass is somewhat arbitrary. The values taken here are for CdS : $m_e = 0.18$, $m_h = 0.53$ and for ZnS : $m_e = 0.42$, $m_h = 0.61$. Eq. (4.81) overestimates the calculated gap shift by more than a factor of two for very small radii ($d < a_B$). Unfortunately, it has not been analyzed, how much of this discrepancy is due to the nonparabolicity of the energy bands in the bulk material and how much is caused by the confinement induced band mixing.

The comparison with some experimental data for CdS is shown in Fig. 4.27. The experimental points are halfway between the predictions of the tight binding calculations and those of the elementary confinement theory.

Chapter 5
OPTICAL PROPERTIES OF SMALL DOTS

In this and the following chapter we analyze important aspects of the optical properties of quantum dots. We treat large and small dots in separate chapters since quite different approaches and approximations are useful for these size regimes. First, in the present Chap. 5 we discuss the optical properties of quantum dots whose sizes are comparable to the bulk exciton Bohr radius. The case of large quantum dots is the subject of the following Chap. 6.

We begin our analysis of optical properties in Sec. 5-1 with an outline of the dipole approximation for the case of quantum dots interacting with a transverse electromagnetic wave. Most calculations for small quantum dots are based on the knowledge of the eigenvalues and wavefunctions of the one- and two- electron-hole-pair states. Using these results, we calculate the linear and third-order nonlinear optical properties. First we present an overview of linear optical properties in Sec. 5-2, and then we deal with the third-order optical nonlinearities in Secs. 5-3 and 5-4.

The important aspect to deal with the optical nonlinearities is to introduce explicitly the irradiating optical field and thus to avoid supplementary assumptions about the induced statistical state of the system. Such an approach is always possible in the framework of perturbation theory with respect to the applied field. In Sec. 5-3 we present the formal derivation of the third-order susceptibility, χ_3, and then we apply the results to compute the nonlinear absorption of small quantum dots (Sec. 5-4). We show that the nonlinear optical response of these quantum dots is dominantly determined by the saturation of the one-pair (exciton) states and the induced (increasing) absorption due to the two-pair (biexciton) states. The influence of the homogeneous broadening and of surface polarization effects is outlined.

In Sec. 5-5 we analyze the dynamic optical response of quantum dot systems. We discuss the phenomenon of photon echo spectroscopy as a representative example of the large class of time resolved four-wave mixing phenomena. The calculations show photon echoes due to the inhomogeneous broadening caused by the size distribution of the dots. The echo signals exhibit clear quantum-beat signatures due interferences between signals arising from the different quantum-confined levels.

Finally, in Sec. 5-6 we summarize recent results of two-photon spectroscopy in quantum dots. Due to their different selection rules, comparison of experimental results for one- and two-photon transitions allows to obtain information about the electron-hole-pair states in the dots.

5-1. Dipole Interaction

We want to treat the interaction of the electrons and holes in the quantum dot with an applied optical field $\mathbf{E}(t)$ in the frame of the dipole approximation

$$\mathcal{H}_{int}(t) = - P\ E(t)\ , \tag{5.1}$$

where

$$\mathbf{E}(t) = E(t)\ \mathbf{n}\ . \tag{5.2}$$

The quantity P is the projection of the dipole moment operator on the direction of the field

$$P = \mathbf{n} \cdot \mathbf{P}\ . \tag{5.3}$$

The form of the dipole interaction in Eq. (5.1) originates from the term

$$\mathcal{H}_{int}(t) = \int d^3r\ \mathbf{j}(\mathbf{r}) \cdot \mathbf{A}(\mathbf{r}, t) \tag{5.4}$$

of the true electromagnetic interaction with a transverse electrical field

$$\mathbf{E} = - \partial/\partial t\ \mathbf{A}\ ,\quad \nabla \cdot \mathbf{A} = 0\ , \tag{5.5}$$

neglecting the coordinate dependence of the field (long wavelength approximation). For a periodic field,

$$\mathbf{E}(t) = \mathbf{n}\ E\ \cos(\omega t)\ , \tag{5.6}$$

the interaction Hamiltonian can be written in the form

$$\mathcal{H}_{int}(t) = \frac{E}{\omega} \sin(\omega t) \int d^3r \; \mathbf{j}(\mathbf{r}) \cdot \mathbf{n} \; . \tag{5.7}$$

The current-density operator in the quantum dot is constructed in the usual way,

$$\mathbf{j}(\mathbf{r}) = -i \; \frac{e\hbar}{2m} \; \Psi^{\dagger}(\mathbf{r}) \; \nabla \; \Psi(\mathbf{r}) + \text{h.c.} \; . \tag{5.8}$$

Here, $\Psi(\mathbf{r})$ and $\Psi^{\dagger}(\mathbf{r})$ are the annihilation and creation operators, and m is the free electron mass. In the effective mass envelope function approximation for quantum confined systems, the one-particle wavefunctions of the system are given as a product of a Bloch part and an envelope part. The Bloch part is taken as the wavefunction of the electron in the bulk crystal (at $k = 0$), and the envelope function is determined by the boundary conditions. For the case of quantum dots we have

$$\Psi(\mathbf{r}) = u_c(\mathbf{r}) \sum_e \phi_e(\mathbf{r}) \, a_e + u_v(\mathbf{r}) \sum_h \phi_h^*(\mathbf{r}) \, a_h^{\dagger} \; . \tag{5.9}$$

The matrix elements which appear through the substitution of Eq. (5.9) into Eq. (5.7) are

$$\int d^3r \; u_c^*(\mathbf{r}) \; \phi_e^*(\mathbf{r}) \; \nabla \; [u_c(\mathbf{r}) \, \phi_{e'}(\mathbf{r})] \tag{5.10a}$$

for the transitions between conduction bands,

$$\int d^3r \; u_v^*(\mathbf{r}) \; \phi_h(\mathbf{r}) \; \nabla \; [u_v(\mathbf{r}) \, \phi_{h'}^*(\mathbf{r})] \tag{5.10b}$$

for the transition between valence bands, and

$$\int d^3r \; u_c^*(\mathbf{r}) \; \phi_e^*(\mathbf{r}) \; \nabla \; [u_v(\mathbf{r}) \; \phi_h(\mathbf{r})] \tag{5.10c}$$

for the valence to conduction band transitions, respectively. Since the envelope functions are usually considered to be constant within an elementary cell, one can separate the integrals into integrations of the Bloch parts over an elementary cell (of volume v) and an integration of the envelope parts over the dot,

$$\int d^3r \simeq \int d^3r_{envelope} \; \frac{1}{v} \int_v d^3r_{Bloch} \; . \tag{5.11}$$

Using the product rule, we can write the gradient terms in (5.10) as

$$\nabla \; [u_i(\mathbf{r}) \; \phi_j(\mathbf{r})] = \nabla \; [u_i(\mathbf{r})] \; \phi_j(\mathbf{r}) + u_i(\mathbf{r}) \; \nabla \; \phi_j(\mathbf{r}) \; . \tag{5.12}$$

For each one of the different transition matrix elements (5.10), only one of the two terms on the RHS of (5.12) survives. Due to the orthonormality of the Bloch parts and of the envelope parts, only the second term on the RHS of (5.12) contributes for the intraband transitions between conduction or valence bands, (5.10a) and (5.10b). In contrast for interband transitions between valence and conduction band, (5.10c), the first term on the RHS of (5.12) contributes, i.e. the term where the gradient acts on the Bloch part.

Summarizing these results, we obtain the dipole operator for intraband transitions (for example conduction–conduction) as

$$-i \; \frac{e\hbar v}{m\omega} \sum_{e,\,e'} a_e^\dagger \; a_{e'} \int d^3r \; \phi_e^*(\mathbf{r}) \; \mathbf{n} \cdot \nabla \; \phi_{e'}(\mathbf{r}) \tag{5.13}$$

and for the interband transitions as

$$-i \; \frac{e\hbar v}{m\omega} \sum_{e,\,h} a_e^\dagger \; a_h^\dagger \int d^3r \; u_c^*(\mathbf{x}) \; \mathbf{n} \cdot [\nabla \; u_v(\mathbf{r})] \int d^3\mathbf{r} \; \phi_e^*(\mathbf{r}) \; \phi_h(\mathbf{r}) \; . \tag{5.14}$$

The matrix elements of the gradient operator may be expressed through the matrix elements of the coordinate operator if one uses the identity

$$-i\hbar\nabla = \frac{m}{i\hbar}\left[-\frac{1}{2m}\hbar^2\nabla^2 + V(\mathbf{r}) , \mathbf{r}\right],$$ (5.15)

which is valid for any mass m and any potential $V(\mathbf{r})$. Then

$$-i\hbar\int d^3r \, \phi_e^*(\mathbf{r}) \, \mathbf{n}\cdot\nabla \, \phi_{e'}(\mathbf{r}) = m_e \frac{\epsilon_e - \epsilon_{e'}}{i\hbar}\int d^3r \, \phi_e^*(\mathbf{r}) \, (\mathbf{n}\cdot\mathbf{r}) \, \phi_{e'}(\mathbf{r})$$ (5.16)

and

$$\int d^3r \, u_c^*(\mathbf{r}) \, \mathbf{n}\cdot\nabla \, u_h(\mathbf{r}) = m \frac{E_c - E_v}{i\hbar}\int d^3r \, u_c^*(\mathbf{r}) \, \mathbf{n}\cdot\mathbf{r} \, u_v(\mathbf{r}) .$$ (5.17)

Here we denoted by ϵ_e the eigenenergies of the envelope functions and took into account that the envelope part of the kinetic energy contains the effective mass m_e, whereas the eigenenergy of the Bloch part is the conduction (valence) band edge, E_c (E_v), and its kinetic energy contains the free electron mass m. Often in applications the energy differences are set equal to the photon energy $\hbar\omega$. Eqs. (5.16) and (5.17) show that the angular momentum change $\Delta\ell$ involves the envelope functions (Bloch functions) for intraband (interband) transitions. Consequently, the *angular momentum change for the envelope functions* has to be

$$\Delta\ell = 1$$ (5.18a)

for intraband transitions and

$$\Delta\ell = 0$$ (5.18b)

for interband transitions, respectively.

As final remark at this point we want to mention that after all these approximations the dipole interaction Hamiltonian actually is

$$E\sin(\omega t) \, i \, [\, P^{(+)} - P^{(-)} \,]$$

instead of the form

$$E \cos(\omega t) \left[P^{(+)} + P^{(-)} \right]$$

which is implied by Eq. (5.1) and which is preferred in applications. However, it can be shown that both forms are physically equivalent.

5-2. Linear Optical Properties

Schematically, the energy level scheme for the states with zero, one, and two-electron-hole pairs is shown in Fig. 5.1, where o, e, and b denotes

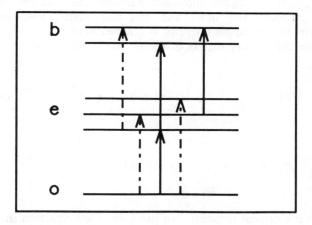

Fig. 5.1: Sketch of the energy level scheme in quantum dots showing the states with zero, one, and two-electron-hole pairs. Here, o, e, and b denote the vacuum, exciton, and the biexciton states, respectively. The solid arrows indicate strong dipole transitions and the dashed ones indicate weak dipole transition induced by the Coulomb interaction.

the vacuum, exciton, and the biexciton states, respectively. Using the projection operator $\Pi_{ij} = |i\rangle \langle j|$, where i and j can be o, e, or b, the Hamiltonian of the quantum dot system can be written as ($\hbar \equiv 1$),

$$\mathcal{H} = \sum_e \omega_e \, \Pi_{ee} + \sum_b \omega_b \, \Pi_{bb}$$

$$- \sum_e \left[d_{eo} \, E(t) \, \Pi_{eo} + \sum_b d_{be} \, E(t) \, \Pi_{be} + \text{h.c.} \right] . \qquad (5.19)$$

Here ω_e and ω_b are the eigenenergies of the one- and two-pair states, and d_{eo} and d_{be} are the dipole moments between the vacuum and one-pair states and between the one- and two-pair states, respectively. (The intraband part of the dipole operator is ignored here.)

We write the density matrix of the quantum dot model as

$$\rho = \rho_{oo} \, |o\rangle \langle o| + \sum_e \left[\rho_{eo} \, |e\rangle \langle o| + \text{h.c.} \right] + \sum_b \left[\rho_{bo} \, |b\rangle \langle o| + \text{h.c.} \right]$$

$$+ \sum_{ee'} \rho_{ee'} |e\rangle \langle e'| + \sum_{bb'} \rho_{bb'} \, |b\rangle \langle b'|. \qquad (5.20)$$

To compute the optical response, we use Liouville's equation in the form

$$\frac{\partial}{\partial t} \rho = -\frac{i}{\hbar} [\, \mathcal{H} \,, \rho \,] - \frac{1}{\hbar} \mathcal{K}\{\rho\} . \qquad (5.21)$$

where $\mathcal{K}\{\rho\}$ models all dissipative processes. Computing the reversible part of the dynamics yields the following set of equations for the expansion coefficients,

$$i \, \frac{\partial}{\partial t} \rho_{ee'} = (\omega_e - \omega_{e'}) \, \rho_{ee'} + (d_{oe'} E \, \rho_{eo} - d_{eo} E^* \rho_{oe'})$$

$$- \sum_b (d_{eb} E \rho_{be'} - d_{be'} E^* \rho_{eb}) , \qquad (5.22)$$

$$i \frac{\partial}{\partial t} \rho_{bb'} = (\omega_b - \omega_{b'}) \rho_{bb'} + \sum_e (d_{eb'} E^* \rho_{be} - d_{be} E \rho_{eb'}) \ , \quad (5.23)$$

$$i \frac{\partial}{\partial t} \rho_{eo} = \omega_e \, \rho_{eo} + \sum_{e'} d_{e'o} E \rho_{ee'} - \sum_b d_{eb} E^* \rho_{bo} - d_{eo} E \rho_{oo} \ , \quad (5.24)$$

$$i \frac{\partial}{\partial t} \rho_{be} = (\omega_b - \omega_e) \rho_{be} - \sum_{e'} d_{be'} E \rho_{e'e} - \sum_{b'} d_{b'e} E \rho_{bb'} + d_{oe} E^* \rho_{bo} \ ,$$

$$(5.25)$$

$$i \frac{\partial}{\partial t} \rho_{bo} = (\omega_b - \omega_e) \rho_{bo} - \sum_e (d_{be} E \rho_{eo} - d_{eo} \, E \rho_{be}) \ . \quad (5.26)$$

For compactness of notation we neglected here all incoherent processes, but in the numerical evaluations we later include the dissipation by adding the proper damping rates.

We evaluate the coupled set of dynamic equations, (5.22) - (5.26), in Sec. 5-5 for the case of a two-pulse photon-echo configuration. In this section we are only interested in the linear optical properties, i.e., in the linear response to the harmonic probe field

$$E(t) = E_p \, e^{i\omega t} \ . \quad (5.27)$$

To evaluate the first-order complex susceptibility,

$$\chi_1(\omega) = \lim_{E_p \to 0} P/E_p \ , \quad (5.28)$$

we need the linear term in the expression of the polarization

$$P = \sum_e d_{eo} \, \rho_{oe} + \text{h.c.} \ . \quad (5.29)$$

Evaluating ρ_{oe} linear in E_p yields

$$\chi_1(\omega) = \frac{i}{\hbar} \sum_e |d_{oe}|^2 \left[\frac{1}{\gamma_e + i(\omega_e - \omega)} + \frac{1}{\gamma_e - i(\omega_e + \omega)} \right]. \qquad (5.30)$$

A detailed discussion of dissipative processes is given in the following Sec. 5-3. Here we modelled dissipation simply as

$$\langle o | \mathcal{K}(\rho) | e \rangle = \hbar \gamma_e \, \rho_{oe} \,, \qquad (5.31)$$

describing the decay of ρ_{oe}. All other matrix elements of $\mathcal{K}$ have been neglected since they do not enter into the final expression of the linear susceptibility. Eq. (5.30) was used without derivation already in Sec. 2-2.

Taking the imaginary part of Eq. (5.30) yields the linear absorption coefficient

$$\alpha(\omega) = \frac{4\pi\omega}{\hbar c \sqrt{\epsilon_2}} \sum_e |d_{oe}|^2 \frac{\gamma_e}{\gamma_e^2 + (\omega_e - \omega)^2} \qquad (5.32)$$

where only the resonant part was taken into account. This result shows that the absorption spectrum of a single quantum dot consists of a series of Lorentzian peaks centered around the one-electron-hole pair energies $\hbar\omega_e$. To obtain the absorption of a distribution of quantum dots, one has to include the size averaging discussed in Sec. 2-3. Examples of the results are plotted in Figs. 2.4 and 2.5.

To study the influence of the valence-band mixing on the linear absorption spectra [Koch et al. (1992)], we use the computed energy levels and the dipole matrix elements obtained from the calculations in Sec. 4-5. To have allowed one-photon transitions, we have to consider states which satisfy the selection rules due to the conservation of angular momentum. Since the angular momentum change of unity resulting from the application of the dipole operator is taken over by the Bloch part of the wavefunctions (dipole matrix element p_{cv}), the total orbital angular momentum of the envelope function of the created electron-hole state has to be zero, Eq. (5.18b). Some of the additional complications arising in the case with spin-orbit coupling have been outlined in Sec. 4-5.

Examples of the computed linear absorption spectra are shown in Fig. 5.2. The energies and dipole matrix elements for these calculations were obtained on the basis of the matrix-diagonalization results including band-mixing effects as outlined in Sec. 4-5. The noticeable feature of these

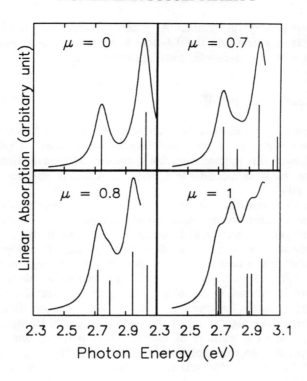

Fig. 5.2: Computed linear absorption spectra for semiconductor quantum dot and various heavy- and light-hole coupling constants μ. The energy band gap has been chosen as E_g = 2.58 eV, and the binding energy of the exciton has been taken as 27 meV, both typical for bulk CdS. The dot radius $R = a_B$ and the mass ratio $m_e \gamma_1 / m_0 = 0.5$. The broadening of the spectra has been taken as $\gamma = 2E_R$. The inserted lines indicate the oscillator strengths for the dipole transitions. From Koch *et al.* (1992).

spectra is that with increasing effective-mass parameter μ a shoulder appears on the high energy side of the energetically lowest transition. This is due to the fact that the coupling between the light and heavy hole opens new channels for dipole transitions [Xia (1989)]. In the parabolic band approximation, the ground state exciton contains mainly the electron-hole-pair wavefunction $(n_e, n_h) = (1s, 1s)$, and the transition to the state $(n_e, n_h) = (1s, 1d)$ is dipole forbidden. However, when the heavy-hole to light-hole coupling strength μ is large the coupling between holes with $1d$ orbit and

1s orbit becomes very important. As a consequence, the dipole transition between the 1s-electron and the 1d-hole becomes dipole allowed, in contrast to the results of a parabolic band approximation.

5-3. Nonlinear Susceptibility

In this section we discuss the technical details related to the theory of nonlinear susceptibilities in semiconductor quantum dots. This concept is well applicable to this system, because quantum dots have discrete energy levels, unlike bulk semiconductors with continuous spectra [Haug and Koch (1990, 1993)].

In order to obtain the proper dependence of the resulting susceptibilities on system volume and damping constants, we first review the relevant aspects of the general nonlinear response theory. As in the previous section we study a system which is described by the Hamiltonian $\mathcal{H}$ and which interacts with a classical optical electric field through the time-dependent interaction Hamiltonian in dipole approximation according to Eq. (5.1). We are interested in calculating the quantum statistical average of the polarization (dipole) operator P. Of course, P has no diagonal matrix elements between the eigenstates of $\mathcal{H}$ since the optical transitions in semiconductors change the number of electron-hole pairs by one, whereas $\mathcal{H}$ commutes with the electron-hole-pair number operator. Consequently, the average of P vanishes in thermal equilibrium. A non-vanishing polarization can only be induced by an electromagnetic field $E(t)$.

Introducing the field at time $t = t_0$, we expand P in powers of the applied field as

$$\frac{1}{\Omega} \langle P(t) \rangle = \int_{t_0}^{t} dt_1 \, \chi_1(t, t_1) \, E(t_1)$$

$$+ \int_{t_0}^{t} dt_1 \int_{t_0}^{t_1} dt_2 \int_{t_0}^{t_2} dt_3 \, \chi_3(t, t_1, t_2, t_3) \, E(t_1) \, E(t_2) \, E(t_3) + \ldots$$

$$(5.33)$$

[Bloembergen (1965)], where we introduced the volume of the system Ω for a correct normalization of the susceptibilities. Terms with even powers of E are excluded, since we assume a system with inversion symmetry and P has the same space-reflection property as the field E.

We write the Liouville equation (5.21) in the form

$$i\hbar \frac{\partial}{\partial t} \rho = [\mathcal{H} + \mathcal{H}_{int}(t), \rho] - i \mathcal{K}\{\rho\} \tag{5.34}$$

with the initial equilibrium condition

$$\rho(t_0) = \rho_0(\mathcal{H}) . \tag{5.35}$$

The "super-operator" $\mathcal{K}$, which acts directly on the matrix elements of the density matrix, represents all the dissipative irreversible processes which assure the existence of asymptotically stationary solutions for harmonic fields. A simple phenomenological "super-operator" which conserves the hermiticity, trace and positivity of the density matrix, and which brings the system back to equilibrium in the absence of the perturbation, is given by [Lindblad (1976)]

$$\mathcal{K}\{\rho\} = \frac{1}{2} \{ \Gamma, \rho\} - \frac{1}{2} \text{tr}(\Gamma\rho_0) [\Gamma, \rho_0] + \text{tr}(\Gamma\rho) , \tag{5.36}$$

where $\{ , \}$ is the anti-commutator and Γ is a positive (hermitian) operator. Eq. (5.36) leads to a generalization of the simple formula

$$\mathcal{K}\{\rho\} = \frac{\hbar}{\tau} (\rho - \rho_0) \tag{5.37}$$

which is obtained if Γ is taken to be just the constant $\hbar/\tau$. Therefore $\mathcal{K}/\hbar$ can be interpreted as the inverse relaxation time operator. In what follows we shall restrict ourselves to operators Γ which commute with the Hamiltonian $\mathcal{H}$.

Formally, one often derives expressions for the optical susceptibility through straightforward application of time-dependent adiabatic perturbation theory, introducing an adiabatic factor $\exp(t/\tau)$ in the field. However, the result coincides with the introduction of a relaxation time τ in the Liouville equation only for the linear susceptibility. For the (third-order) nonlinear susceptibility this procedure gives wrong coefficients for the different terms and for the damping parts of the denominators [Bányai et al. (1988b)]. Beyond linear response, the addition of an ideal relaxation term in the Liouville equation or the adiabatical introduction of the perturbation are physically and mathematically different. In the nonlinear theory one therefore has to be very careful and should avoid to introduce arbitrary damping in an adiabatically obtained formula. A consistent treatment is assured using the Liouville equation in the form of Eq. (5.34) since the damping mechanisms described by dissipative super-operators assure that all the important general physical properties of the original quantum mechanical system are conserved.

After these remarks we now proceed to obtain the formal solution of the generalized Liouville equation (5.34). To simplify the notations we introduce the super-operator (Liouvilleian) $\mathscr{L}$ as

$$\mathscr{L}\{\rho\} = -\frac{i}{\hbar} [\, \mathscr{H} \,,\, \rho \,] - \frac{1}{\hbar} \mathscr{K}\{\rho\} \,. \tag{5.38}$$

It is convenient to express Eqs. (5.34) – (5.35) through the equivalent integral equation

$$\rho(t) = \rho_0 + \frac{1}{i\hbar} \int_{t_0}^{t} dt' \, \exp[\mathscr{L}(t-t')] \Big[P \,,\, \exp[\mathscr{L}(t'-t)] \, \rho(t') \Big] E(t') \,. \tag{5.39}$$

This form is useful for iterative solutions of ρ in powers of the field E. We avoid here the mathematical details and give directly the results of the first and third iterations for the susceptibilities. They are expressed in terms of the transposed Liouvilleian $\mathscr{L}^T$ defined through the identity

$$\mathrm{tr}\Big[A \, \mathscr{L}\{B\} \Big] = \mathrm{tr}\Big[B \, \mathscr{L}^T\{A\} \Big] \,, \tag{5.40}$$

where A and B are arbitrary operators. We denote the equilibrium average as

$$\langle A \rangle_0 = \mathrm{tr}\Big[\rho_0 \, A \Big] \,. \tag{5.41}$$

With these notations we have

$$\chi_1(t,t_1) = -\frac{1}{i\hbar\Omega} \Big\langle \Big[P, \exp[\mathscr{L}^T(t-t_1)] \, P \Big] \Big\rangle_0 \tag{5.42}$$

and

$$\chi_3(t, t_1, t_2, t_3) = - \frac{1}{\Omega(i\hbar)^3}$$

$$\left\langle \left[P, \exp[\mathcal{L}^T (2t_2 - t_1 - t_3)] \right] \left[P, \exp[(\mathcal{L}^T (2t_1 - t - t_2)] \left[P, \exp[\mathcal{L}^T (t - t_1)] P \right] \right] \right\rangle_0,$$

(5.43)

where we took into account that $L\{\rho_0\} = 0$.

It is important to notice that the susceptibilities are invariant with respect to a common choice of the time origin. Therefore, we have only three independent time variables and it is useful to define the susceptibility as

$$\chi_3(\tau_1, \tau_2, \tau_3) \equiv \chi_3(t, t-\tau_1, t-\tau_1-\tau_2, t-\tau_1-\tau_2-\tau_3) \ .$$

(5.44)

Furthermore, we set $t_0 = -\infty$, which does not affect the expressions since $E(t)$ is vanishing for $t < t_0$ and τ_1, τ_2, τ_3 are all positive.

The important steps in the evaluation of the time-dependent susceptibilities are the introduction of the complete set of eigenstates $|n\rangle$ of $\mathcal{H}$ as the intermediary states ($\Sigma_n |n\rangle \langle n| = 1$) and the use of diagonal damping Γ, which enables us to write

$$\langle n| \mathcal{L}^T \{A\} |n'\rangle = \frac{i}{\hbar} \left(E_n - E_{n'} + i \frac{\Gamma_n + \Gamma_{n'}}{2} \right) A_{nn'} + \frac{1}{\hbar} \delta_{nn'} \Gamma_n \frac{\langle \Gamma A \rangle_0}{\langle \Gamma \rangle_0},$$

(5.45)

where A is again an arbitrary operator. In particular for operators for which $\langle \Gamma A \rangle_0 = 0$, the action of the super-operator $\mathcal{L}^T$ can be evaluated easily. Utilizing the positivity of Γ we can write generally

$$\mathcal{L}^T \{A\} = \mathcal{L}^T \left\{ A - \frac{1}{\Gamma} \langle \Gamma A \rangle_0 \right\} + \mathcal{L}^T \left\{ \frac{1}{\Gamma} \right\} \langle \Gamma A \rangle_0$$

$$= \frac{1}{\hbar} \left(i \left[\mathcal{H}, A - \frac{1}{\Gamma} \langle \Gamma A \rangle_0 \right] - \left[\Gamma, A - \frac{1}{\Gamma} \langle \Gamma A \rangle_0 \right]_+ + \left(\frac{\Gamma}{\langle \Gamma \rangle_0} - 1 \right) \langle \Gamma A \rangle_0 \right).$$

(5.46)

This equation allows to use the same decomposition trick for the super-operator $\mathcal{L}^T$ in the exponent,

$$\exp(t \, \mathcal{L}^T) \, \{A\} = \exp\left[\frac{(i\mathcal{H} - \Gamma)t}{\hbar}\right] \left(A - \frac{\langle \Gamma A \rangle_0}{\Gamma}\right)$$

$$\times \exp\left[-\frac{(i\mathcal{H} + \Gamma)t}{\hbar}\right] + \langle \Gamma A \rangle_0 \exp(t \, \mathcal{L}^T) \left\{\frac{1}{\Gamma}\right\}. \qquad (5.47)$$

The exponential of the super-operator applied to $1/\Gamma$ could be treated similarly. However, it turns out that these terms can be eliminated from the final expression through other tricks. Special care has to be devoted to the time integration. The exponents occur in a form such that only the negative (decaying at $-\infty$) exponentials survive, assuring the convergence of all integrations.

The time-dependent third-order susceptibility is important in the treatment of experiments with very short pulses. On the other hand, for the discussion of quasi-stationary light fields (very long laser pulses) a useful entity is obtained by introducing the Fourier transforms,

$$\frac{1}{\Omega} \langle P(\omega) \rangle = \chi_1(\omega) \, E(\omega)$$

$$+ \int d\omega_1 \, d\omega_2 \, d\omega_3 \, \chi_3(\omega_1, \omega_2, \omega_3) \, \delta(\omega - \omega_1 - \omega_2 - \omega_3) \, E(\omega_1) \, E(\omega_2) \, E(\omega_3) + \dots$$

$$(5.48)$$

The expression of $\chi_3(\omega_1, \omega_2, \omega_3)$ is obtained as

$$\chi_3(\omega_1, \omega_2, \omega_3) = - \frac{1}{\Omega(i\hbar)^3} \left\langle \left[P, \frac{1}{\mathcal{L}^T + \omega_1 + \omega_2 + \omega_3} \left[P, \frac{1}{\mathcal{L}^T + \omega_2 + \omega_3} \left[P, \frac{1}{\mathcal{L}^T + \omega_3} P\right]\right]\right] \right\rangle_0$$

$$(5.49)$$

Actually, only the symmetrical part $\chi_3^{sym}(\omega_1, \omega_2, \omega_3)$ of $\chi_3(\omega_1, \omega_2, \omega_3)$ contributes to the final result.

A substantial variety of experimental situations with beams of different frequencies may be discussed in terms of the third-order susceptibilities. For example in a one-beam experiment the quantity

$$3 \, \chi_3^{sym}(\omega, \omega, -\omega)$$

is relevant, whereas in a pump and probe setting the term

$$6 \, \chi_3^{sym}(\omega_P, -\omega_P, \omega_T)$$

appears, where ω_P, ω_T are the frequencies of the pump and probe fields respectively. The explicit form of the frequency-dependent susceptibility is obtained again through the introduction of a complete set of eigenstates of the Hamiltonian $\mathscr{H}$ as the intermediate states.

The delicate point of the infinite volume limit in the presence of a continuous spectrum is the fact, that although the whole susceptibility is an intensive quantity, the decomposition through the use of intermediate states formally introduces a number of terms which are proportional to the volume or even the volume squared. To correctly eliminate these artificial volume dependencies, one has to be aware of a number of quite subtle cancellation mechanisms.

Let us now specify the results for the pump-probe configuration in a semiconductor quantum dot. For the experimentally relevant case of band-edge spectroscopy we retain here only the interband part of the polarization operator. It has nonvanishing matrix elements P_{oe} between the vacuum and one-pair (exciton) states as well as P_{eb} between the one-pair and two-pair (biexciton) states. The explicit expression of the pump-probe nonlinear susceptibility has been derived by Hu *et al.* (1990b),

$$\chi_3^{sym}(\omega_P, -\omega_P, \omega_T) = -\frac{i}{6\Omega}[T_1(\omega_P, -\omega_P, \omega_T) + T_2(\omega_P, -\omega_P, \omega_T)] \quad (5.50)$$

where

$$T_1 = \sum_{ee'} \frac{|P_{eo}|^2 \, |P_{e'o}|^2}{\Gamma_{eo} + i(\epsilon_e - \hbar\omega_T)} \left[\frac{2\Gamma_{e'o}/\Gamma_{e'e'}}{\Gamma_{e'o}^2 + (\epsilon_{e'} - \hbar\omega_P)^2} \right.$$

$$+ \left(\frac{1}{\Gamma_{e'o} - i(\epsilon_{e'} - \hbar\omega_P)} + \frac{1}{\Gamma_{e'o} + i(\epsilon_{e'} - \hbar\omega_T)} \right) \frac{1}{\Gamma_{e'e'} - i(\hbar\omega_T - \hbar\omega_P)}$$

$$+ \frac{1}{\Gamma_{e'o} - i(\epsilon_{e'} - \hbar\omega_P)} \left(\frac{1}{\Gamma_{ee'} + i\epsilon_{ee'}} + \frac{1}{\Gamma_{ee'} + i(\epsilon_{ee'} - \hbar\omega_T + \hbar\omega_P)} \right)$$

$$+ \frac{1}{\Gamma_{eo} + i(\epsilon_e - \hbar\omega_P)} \frac{1}{\Gamma_{ee'} + i\epsilon_{ee'}} + \frac{1}{\Gamma_{eo} + i(\epsilon_e - \hbar\omega_T)} \frac{1}{\Gamma_{ee'} + i(\epsilon_{ee'} - \hbar\omega_T + \hbar\omega_P)} \right]$$

$$(5.50a)$$

and

$$T_2 = \sum_{bee'} P_{oe}\, P_{eb}\, P_{be'}\, P_{e'o} \left[\frac{1}{\Gamma_{bo}+i(\epsilon_b -\hbar\omega_P -\hbar\omega_T)} \left(\frac{1}{\Gamma_{eo}+i(\epsilon_e -\hbar\omega_T)} \right. \right.$$

$$\left. - \frac{1}{\Gamma_{be}+i(\epsilon_{be}-\hbar\omega_T)} \right) \left(\frac{1}{\Gamma_{e'o}+i(\epsilon_{e'}-\hbar\omega_P)} + \frac{1}{\Gamma_{e'o}+i(\epsilon_{e'}-\hbar\omega_T)} \right)$$

$$- \frac{1}{\Gamma_{be}+i(\epsilon_{be}-\hbar\omega_T)} \left(\frac{1}{\Gamma_{eo}-i(\epsilon_e -\hbar\omega_P)} \left(\frac{1}{\Gamma_{ee'}-i\epsilon_{ee'}} + \frac{1}{\Gamma_{ee'}-i(\epsilon_{ee'}+\hbar\omega_T - \hbar\omega_P)} \right) \right)$$

$$+ \frac{1}{\Gamma_{e'o}+i(\epsilon_{e'}-\hbar\omega_P)} \frac{1}{\Gamma_{ee'}-i\epsilon_{ee'}} + \frac{1}{\Gamma_{e'o}+i(\epsilon_{e'}-\hbar\omega_T)} \frac{1}{\Gamma_{ee'}-i(\epsilon_{ee'}+\hbar\omega_T - \hbar\omega_P)} \right)\Bigg]$$

$$(5.50b)$$

.

Here, $\epsilon_{ij} = \epsilon_i - \epsilon_j$, and $i = e, b$ runs over the exciton and biexciton states with energies ϵ_i. The damping constants are $\Gamma_{ij} = (\Gamma_i + \Gamma_j)/2$. For $i \neq j$ they represent the phenomenological coherence decay rate of the ij transition while for $i = j$ they describe the population decay of state i. In the following section we apply this result to the cases of small quantum dots. The case of large quantum dots will be analyzed in Chap. 6.

5-4. Third-Order Nonlinearities

To compute the lowest-order optical nonlinearities, we use the third-order susceptibility derived in the previous section. As for the linear optical spectra analyzed in Sec. 5-2 we use the matrix-diagonalization results, but in the present case we need the one- and two-electron-hole-pair states as well as the corresponding dipole matrix elements.

A) Strong Confinement Approximation

Before we discuss the case of interacting electron-hole pairs we briefly summarize some features obtained in the pure strong confinement limit. Here we ignore the Coulomb interaction between the particles and take into account only the lowest lying $\ell = 0$, $m = 0$, $n = 0$ states. Then one has two exciton states with differently oriented spins which may be optically generated,

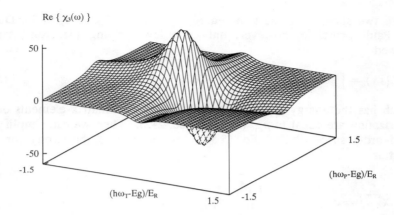

Fig. 5.3a: Strong-confinement approximation. Real part of $\chi_3^{sym}(\omega_P, -\omega_P, \omega_T)$ in units of p_{cv}^4/Ω as function of $\hbar\omega_T$ and $\hbar\omega_P$ (in Rydberg energies) for $R = 0.5\ a_B$ and $\Gamma = 0.3\ E_R$.

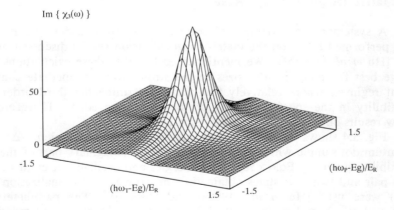

Fig. 5.3b: Imaginary part of $\chi_3^{sym}(\omega_P, -\omega_P, \omega_T)$ for same conditions as Fig. 5.3a.

$$|x\rangle = |\,e000\uparrow, h000\downarrow\,\rangle \quad \text{and} \quad |x'\rangle = |\,e000\downarrow, h000\uparrow\,\rangle \,. \tag{5.51}$$

These two states have the same energy $\epsilon_x = E_G + E_R \, (\pi \, a_B)^2/R^2$. Due to the Pauli principle, however, only one lowest lying biexciton state is allowed

$$|xx\rangle = |\,e000\uparrow, e000\downarrow, h000\uparrow, h000\downarrow\,\rangle \,, \tag{5.52}$$

which has the energy $\epsilon_{xx} = 2\,\epsilon_x$. All the relevant matrix elements of the polarization are equal to p_{cv} (see also Sec. 2-2). Then we can simplify the third-order susceptibility, Eq. (5.50), for the case of the strong confinement as

$$\chi_3^{sym}(\omega_P,-\omega_P,\omega_T) = -\frac{i2p_{cv}^4}{3\Omega}\,\frac{1}{\Gamma+i(\epsilon-\hbar\omega_T)}$$

$$\times\left[\frac{2}{\Gamma^2+(\epsilon-\hbar\omega_P)^2} + \frac{1}{\Gamma+i(\hbar\omega_P-\hbar\omega_T)}\left(\frac{1}{\Gamma-i(\epsilon-\hbar\omega_P)} + \frac{1}{\Gamma+i(\epsilon-\hbar\omega_T)}\right)\right]\,. \tag{5.53}$$

In Fig. 5.3 we show the shape of this complex function of the two frequencies ω_T and ω_P for $R = 0.5 \, a_B$ and $\Gamma = 0.3 \, E_R$.

B) Matrix Diagonalization Results

A systematic investigation of the first- and third-order susceptibility was performed based on the matrix diagonalization results discussed in Sec. 4-2 [Hu *et al.* (1990b)]. We mentioned earlier that these calculations converge best for quantum-dot sizes in the strong and intermediate confinement regimes, where relatively few states determine the third-order susceptibility in the spectral vicinity of the absorption edge. Therefore, we show results only for such relatively small quantum dots.

Fig. 5.4 shows the calculated change in optical transmission, $-\Delta\alpha$, of a quantum dot sample which is proportional to the imaginary part of the susceptibility given by Eq. (5.50). The energies and oscillator strengths of the one-pair and two-pair states obtained from the matrix diagonalization (Sec. 4-2) were used [Hu *et al.* (1990a) and (1990b)]. The parameters are $R/a_0 = 1$, $\epsilon = 1$, and $m_e/m_h = 0.24$. In Figs. 5.4 *a - c* the calculated results are shown for the frequency regime around the lowest exciton resonance E_1 for different damping constants $\hbar\Gamma_{ij} = \gamma$. For simplicity all damping constants in Eq. (5.50) were taken equal. The pump frequency ω_L was chosen to be in resonance with the energetically lowest one-pair state. We

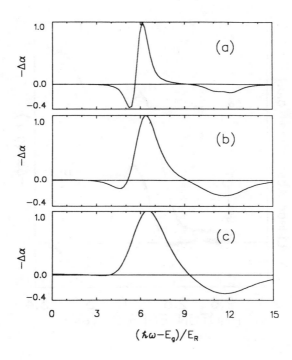

Fig. 5.4: Computed changes of the optical transmission for the parameters $R/a_B = 1$, $\epsilon = 1$, $m_e/m_h = .24$ and three values of the damping: $\gamma/E_R = 1$ (a), 2 (b) and 3 (c). From Hu *et al.* (1990b).

see a positive peak around the pump frequency, indicating the saturation (bleaching) of the one-pair transition. Additionally, we see negative structures on the low and high-energy side of the positive peak. These negative peaks show increasing probe absorption due to the generation of two-electron-hole pair states via absorption of one pump and one probe photon. The resonance on the low-energy side of the positive peak is caused by the ground state biexciton and is known from bulk semiconductors, such as CuCl, which have a large biexciton binding energy. This resonance is visible in quantum dots only for relatively small broadening γ. It is suppressed by the saturating one-pair resonance for increasing γ.

The induced absorption on the high-energy side of the saturating one-pair resonance is caused by transitions to the excited two-pair states dis-

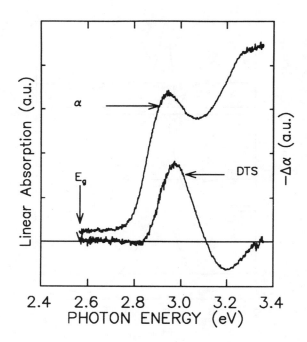

Fig. 5.5: Experimental changes of optical transmission in CdS quantum dots embedded in glass. The linear absorption spectrum as well as the spectral position of the pump frequency are also shown. From Hu *et al.* (1990a).

cussed at the end of Sec. 4-2. As already indicated there, these transitions are possible only since the Coulomb interaction changes the selection rules for dipole transitions. Our analysis shows that the observed transitions occur into those two-pair states, where the two electrons are in their ground states but one or two of the heavier holes are in excited states. Such transitions are dipole allowed only because of the symmetry breaking effect of the Coulomb interaction.

The probe photon generates an electron-hole pair in the presence of the pump-generated pair. Therefore the possibilities for dipole transitions involving the probe photon are different than those of the pump photon. In this sense, the induced absorption resembles the *excited state absorption* in atomic physics. This induced absorption on the high-energy side of the saturating one-pair resonance has been observed in several quantum-dot

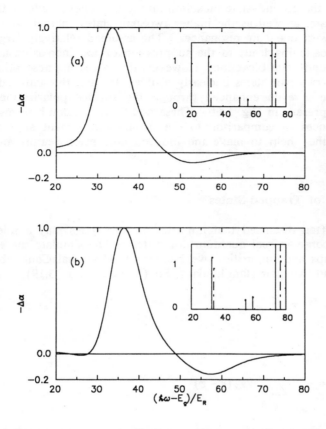

Fig. 5.6: Computed changes of the optical transmission for the parameters $R/a_B = 0.5$, $m_e/m_h = 0.24$, $\gamma/E_R = 10$, $\epsilon = 1$ (a) and $\epsilon = 10$ (b). The oscillator strengths of one- and two-pair transitions are represented in the insets. From Hu et al. (1990b).

samples [Peyghambarian et al. (1989), Park et al. (1990), Morgan et al. (1990)]. In Fig. 5.5 we show an example of such an experimentally measured differential transmission for CdS quantum dots in glass, together with the linear absorption spectrum. The spectral position of the pump is also given.

In Fig. 5.6 the computed [Hu et al. (1990b)] pump-induced transmission changes for different dielectric mismatch between quantum dot and surrounding material are shown. The insets show the one and two-pair dipole transition matrix elements as dashed and full lines, respectively. One clearly sees the ground-state biexciton on the low-energy side of the lowest one-pair state, as well as the higher two-pair states energetically between the two lowest one-pair resonances. The assumed relatively large homogeneous broadening leads to the suppression of the increasing absorption for the ground-state biexciton. However, the increasing absorption due to the excited two pair states is clearly visible. In fact, this induced absorption feature is even enhanced through the surface polarization effects, which are present in Fig. 5.6b. These additional Coulomb terms increase the differences in comparison to the strong-confinement approximation. Therefore they help to make the induced absorption feature more pronounced.

C) Effects of Trapped States

It is often discussed that impurity or trap states play a role for the optical response in real quantum dot systems. To simulate the effects of charged traps and impurities, we include an additional Coulomb term in the one- and two-pair Hamiltonians, Eq. (3.18) and Eq. (3.19),

$$\mathscr{H} \;\rightarrow\; \mathscr{H} + \mathscr{H}_{im} \tag{5.54}$$

where

$$\mathscr{H}_{im} \;=\; -\sum_{i\,=\,e,h} V_{im}(r - z) \tag{5.55}$$

which describes the interaction of the electrons and holes with a single point-charge at position z [Park et $al.$ (1990)]. With this additional term included, we repeat the matrix diagonalization calculations of Sec. 4-2 described above. An example of the resulting linear and nonlinear absorption spectra is shown in Fig. 5.7.

Assuming one positive point charge at the cluster surface, we obtain the results shown in Fig. 5.7. This figure shows that the linear and nonlinear absorption spectra are modified due to the interaction of the photogenerated electron-hole pairs with the additional impurity charge at the surface. However, the overall shape of the spectra is still very similar to those for an intrinsic quantum dots. Also additional calculations, where we assumed a stationary electron and a hole at different positions at the surface yield again similar results. Therefore, we conclude that the qualitative features of the linear and nonlinear absorption spectra in semiconductor quantum dots are not very sensitive to the presence of impurities as long as

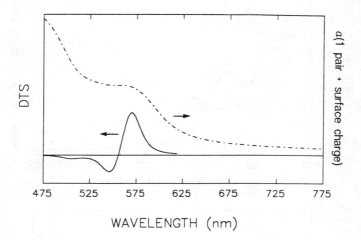

Fig. 5.7: Computed linear absorption spectrum (dashed line) and absorption change (DTS) for pumping into the energetically lowest transition for a semiconductor quantum dots in the presence of an additional point charge at the surface of the dot. From Park *et al.* (1990).

we compute the nonlinear spectra properly using the two electron-hole-pair states in the presence of the additional charges. In fact, the presence of impurities makes it easier to obtain the induced absorption feature at the high-energy side of the saturating one-pair resonance, since it leads to even stronger changes of the dipole transition rules for the probe-generated electron-hole pair.

To investigate a different scenario we show in Fig. 5.8 the comparison of two linear absorption spectra. One spectrum is that of an intrinsic quantum dot and the other one is computed for the case when two electron-hole pairs are present, but one of the holes is trapped at the surface and the other three carriers are free to move inside the dot. Fig. 5.8 shows that the difference between the two linear absorption spectra exhibits a negative peak for low energies (red shift of the lowest transition) and a positive peak on the high-energy side (blue shift of the second lowest transition).

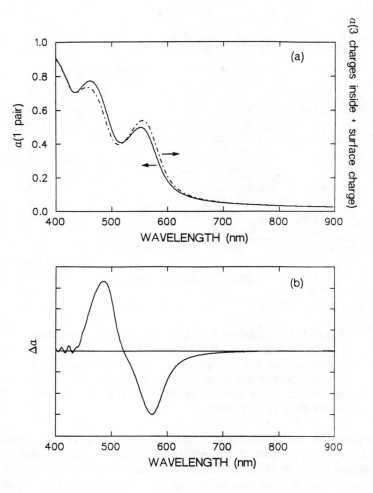

Fig. 5.8: (a) Linear absorption spectra for an intrinsic quantum dot and for a quantum dot with a second electron-hole pair whose hole is trapped at the surface. (b) The absorption change (DTS) shows the difference between the two linear spectra. From Park *et al.* (1990).

5-5. Photon Echo

Time-resolved spectroscopy is a powerful tool to gain valuable insight into the electronic response of materials. In this chapter we discuss one technique, called *time-resolved four wave mixing* or *photon echo spectroscopy*, which is of interest in connection with semiconductor quantum dots due to the inhomogeneous broadening of the optical resonances as a consequence of the dot size distribution.

First, we review briefly the concept behind photon echoes in inhomogeneously broadened systems. If one excites such a system with a pulse at time $t = -t_p$, and then with another pulse at time $t = 0$, the system responds with an echo pulse at the time $t = +t_p$. In atomic photon echo experiments, one uses short-pulse excitation of a system with a broad distribution of transitions, such as a cell of gas in thermal equilibrium where the velocity distribution of the gas atoms or molecules provides a continuum of Doppler shifted transitions.

The physical origin of the photon echo phenomenon is as follows. The first pulse (partially) excites a coherent superposition of many states of the system. Each state develops in time according to its transition energy. The instantaneous state is proportional to $\exp(iE(k)t/\hbar)$; i.e., the different transitions are out of phase. The second pulse (partially) reverses this temporal evolution, which has progressed for the time period t_p between the pulses. Hence, at the time t_p after the second pulse, the states are (partially) back in phase, generating the echo pulse.

The measurement of photon echoes is a special example of the more general class of time-resolved multi-wave mixing experiments. In the conceptually simplest photon echo experiments, the two-pulse echo configuration, the two pulses propagate in different directions which we denote as $\mathbf{k}_x$ and $\mathbf{k}_p$, are sketched in Fig 5.9. The first pulse, E_p, is incident at time $t = -t_p$, and the second pulse, E_x, peaks at $t = 0$. The echo is observed most easily in the direction $2\mathbf{k}_x - \mathbf{k}_p$, the so-called four-wave-mixing or photon-echo direction, where the momentum conservation law is satisfied.

As we discussed in the previous chapters of this book, the blue shift of the quantum-confined electron-hole-pair states and the spacings between the energy levels are approximately proportional to R^{-2}. Due to the strong size dependence of the resonance energies, the size fluctuations in existing quantum dot samples cause a pronounced inhomogeneous broadening. Consequently, dilute quantum dot samples and inhomogeneously broadened atomic systems display some significant similarities.

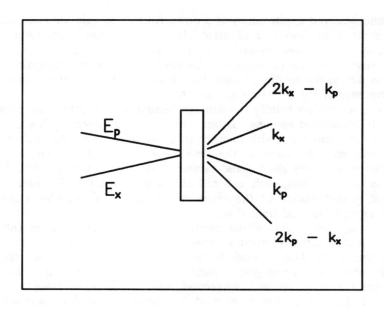

Fig. 5.9: Schematic plot of a four-wave mixing experiment. The pulse E_x (E_p) propagates in the direction k_x (k_p). The transmitted signals are labelled according to their propagation direction as k_x, k_p, $2k_x - k_p$ (four wave mixing or photon echo direction), $2k_p - k_x$ (conjugate direction). The traditional two-beam photon echo is observed in the direction $2k_x - k_p$ at time t_p, if E_x and E_p are pulses and E_x is centered around the time $t = 0$ whereas E_p is incident at $t = -t_p$, respectively.

In atomic systems, photon echo experiments are frequently employed to eliminate Doppler effects and other inhomogeneous broadening mechanisms. Hence, it is not unreasonable to anticipate that photon echo experiments in quantum dot samples can provide valuable information about the quantum confined states which is usually obscured by the large dot size distribution.

In the conceptually simplest photon echo experiments, the two-pulse echo configuration, the two pulses propagate in different directions which we denote as $\mathbf{k}_x$ and $\mathbf{k}_p$. Consequently, the electric field can be written as

$$E(t) = E_x(t)\, e^{i\mathbf{k}_x \cdot \mathbf{R}} + E_p(t)\, e^{i\mathbf{k}_p \cdot \mathbf{R}}, \qquad (5.56)$$

where $\mathbf{R}$ is the macroscopic position vector. The field of Eq. (5.56) induces polarizations propagating along $\mathbf{k}_x$, $\mathbf{k}_p$, $2\mathbf{k}_x - \mathbf{k}_p$, $2\mathbf{k}_p - \mathbf{k}_x$ and higher-order directions. If $E_x \gg E_p$, it is a reasonable approximation to keep only those terms which are linear in E_p and quadratic in E_x. It turns out, that in order to obtain closed equations, we also need the population terms in $\mathbf{k}_x - \mathbf{k}_p$ and $\mathbf{k}_p - \mathbf{k}_x$ directions. The equations governing these polarizations and populations are derived from the spatial Fourier transforms of Eqs. (5.22) – (5.26). The resulting lengthy equations are solved numerically using a fourth order Runge-Kutta method. [Koch $et\ al.$ (1993)]

As discussed above, the size distribution of the quantum dots is the major cause of inhomogeneous broadening in this system. To model this in the calculations, we assume that the dot radii are distributed according to the Gaussian distribution

$$F(R) \propto e^{-(R-R_0)^2/a^2} \qquad (5.57)$$

where R_0 is the average dot size. Although different expressions of the size distribution are possible in quantum dot samples, the specific form does not affect our conclusion in any significant way. In practice, we discretize the continuous distribution (5.57), and compute numerically the energy levels of one- and two-pair states, as well as the dipole moments between the states for each size R_i. Then we compute numerically the signal in the direction $2\mathbf{k}_x - \mathbf{k}_p$ contributed by quantum dots with size R_i. The net echo signal is the coherent summation of the echo signal from the different dots with sizes R_i,

$$P_{echo}(t) \propto \int_0^\infty dR\, F(R)\, P(R,t)_{2k_x - k_p} . \qquad (5.58)$$

For the numerical evaluation we choose the example of CdS quantum dots with perfect confinement conditions. We plot the computed echo signal for various delay times in Fig. 5.10. The inset of Fig. 5.10 shows the computed linear absorption of the same system. Because of the very short dephasing time in existing quantum dot samples, we need ultra-short pulses to resolve the possible echo signals. In this example, we use 20 fs pulses for both pump and probe, where we use the convention that the pump

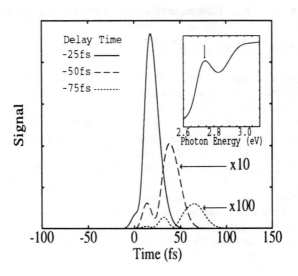

Fig. 5.10: Computed quantum-dot photon-echo signals for pump-probe delay times of 25 fs, 50 fs and 75 fs. The signals for the delay times 50 fs and 75 fs are multiplied by factors of 10 and 100, respectively. The inset shows the computed linear absorption, the central pump and probe pulse energy is indicated by the arrow. The parameters are chosen for CdS, $m_e = 0.2\, m_h$, $E_g = 2.7\ eV$, $E_R = 27$ meV, $T_1 = T_2 = 30\ fs$. The average dot size R_0 is equal to the bulk exciton Bohr radius, the size distribution $a = 0.2\, R_0$. The pump and probe pulses have Gaussian shapes with duration 20 fs (FWHM of intensity) and a maximum field strengths $\mu_{cv}E_2 = 0.1$meV and $\mu_{cv}E_1 = 0.01$ meV, respectively. From Koch et al. (1993).

pulse is incident after the probe pulse. If the duration of the pulses is significantly longer than the dephasing, the polarization induced by the first pulse vanishes before the second pulse arrives. In this case only free induction decay can be observed. The signals in Fig. 5.10 show clear photon echoes at the appropriate delay times. The signal strength decays exponentially as a function of delay time. An interesting feature are the oscillatory structures, which are quantum beats attributed to the interference between the two lowest electron-hole-pair states. Comparing the quantum beat frequencies and the computed energy difference between the lowest quantum dot states for the average size R_0 shows agreement to within ten percent. Even though quantum beats in photon echo signals have been noticed by researchers in atomic physics [Brewer (1977), Stenholm (1977)] the presented calculations nonetheless suggest a novel technique for quantum dot spectroscopy to distinguish the energies of higher quantum confinement states, which are difficult to obtain by conventional pump probe methods.

5-6. Two-Photon Transitions

Information about the band edge states of semiconductor quantum dots can be obtained not only from experiments based on one-photon interband transitions, as always assumed in the previous discussion in this book, but also from multi-photon spectroscopy, the simplest case of which is *two-photon spectroscopy*. Typically, in such a two-photon experiment the crystal is illuminated with two different laser frequencies, which are both chosen in the transparency region of the material to minimize unwanted one-photon absorption. Electron-hole pair excitation is then energetically possible only under the simultaneous absorption of two photons.

Actually, a two-photon contribution is already contained in the third order susceptibility, Eq. (5.50), namely in the terms containing denominators which are resonant for $\epsilon_b - \hbar\omega_T - \hbar\omega_P$. However, these terms describe two-photon transitions where both photon energies exceed the band-gap energy E_g. To obtain the interesting two-photon transition, where $\epsilon_e - \hbar\omega_T - \hbar\omega_P$ gives rise to a resonance, we have to consider also the intraband matrix elements of the polarization operator (see Sec. 5-1) which were ignored in the derivation of Eq. (5.50).

Due to the angular momentum selection rules, the two-photon absorption process generates a state with total angular momentum zero or two, hence probing transitions which are not possible as one-photon transitions with an angular momentum transfer of one. A schematic drawing of the allowed one- and two-photon transitions of this sort is shown in Fig. 5.12.

To compute the optical susceptibility which includes the two-photon transitions we again start from the general operator equation (5.49) for the third order susceptibility, but we take simultaneously the interband and intraband matrix elements of the polarization operator into account. We include only the one-electron-hole-pair states since the relevant photon energies are of the order of $E_{g/2}$, so that the energy denominators implied by two electron-hole-pair states can become very large. Hence, instead of the Hamiltonian (5.19) we take

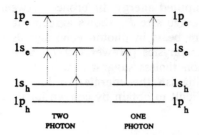

Fig. 5.11: Schematic representation of one- and two-photon excitation in a semiconductor quantum dot. The single-particle electron and hole states are labelled with the quantum numbers appropriate for the strong confinement approximation.

$$\mathcal{H} = \sum_e \omega_e \, \Pi_{ee} - \left[\left(\sum_e d_{eo} \, \Pi_{eo} + \sum_{e,e'} d_{ee'} \Pi_{ee'} \right) E(t) + \text{h.c.} \right].$$

(5.59)

We obtain the two-photon third-order susceptibility as [Hu (1991)]

$$\chi_{2p}^{(3)} = -i|E|^2 \sum_{ee'e''} \frac{d_{oe} d_{ee'} d_{e'e''} d_{e''o}}{i(\omega_{e''} - \omega) + \gamma_{e''o}} \frac{1}{i(\omega_{e'} - 2\omega) + \gamma_{e'o}}$$

$$\times \left[\frac{1}{i(\omega_e - \omega) + \gamma_{eo}} - \frac{1}{i(\omega_{e'} - \omega_e - \omega) + \gamma_{e'e}} \right].$$

(5.60)

Here, we only listed the terms which directly contribute to two-photon transitions. All the other terms, which are related to saturation and transition bleaching, are neglected because the photon energy in this case is approximately one half of that necessary to cause the direct band transitions and no absorption resonances exist at this frequency. The imaginary part of the two-photon optical susceptibility (5.60) yields the two-photon absorption coefficient

$$\alpha_{2p}(\omega) \propto \text{Im} \, [\chi_{2p}^{(3)}(\omega) \,].$$

(5.61)

Eq. (5.61) has been evaluated with and without the spin-orbit coupling effects of Sec. 4-5 [Hu (1991), Kang *et al.* (1992)]. Fig. 5.12 shows an example of the computed two-photon absorption spectrum in comparison to the corresponding one-photon spectrum in the parabolic valence band approximation. We see, that the two-photon peaks occur always at higher

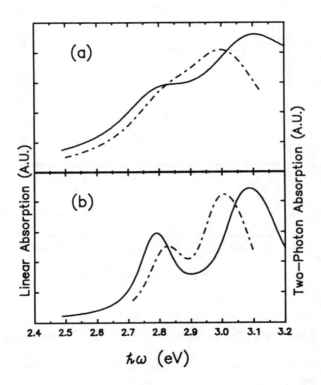

Fig. 5.12: Computed one- (solid) and two-photon (dashed) absorption spectra in the parabolic valence band approximation for for $m_e/m_h = 0.2$, $\epsilon = 1$, $E_g = 2.6$ eV, $E_R = 27$ meV, and the radius of the dot $R = 0.5\, a_B$. The inhomogeneous broadening factor $\delta R/R = 0.15$. The homogeneous broadening constant is $\gamma = 5\, E_R$ in (*a*) and $2\, E_R$ in (*b*). From - Hu (1991).

energies than the one-photon resonances.

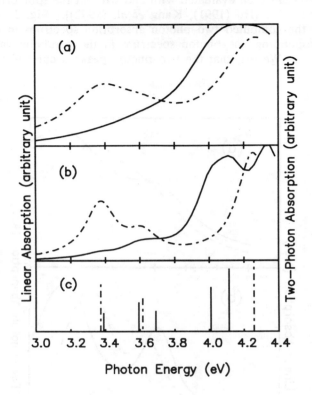

Fig. 5.13: Comparison of one-photon (dashed) and two-photon (solid) absorption spectra computed including valence-band mixing effects. The radius of the dot $R = 0.5a_B$ and coupling strength $\mu = 0.75$. The broadening constant is (a) $\gamma = 8E_R$ and (b) $\gamma = 4E_R$. The other parameters are the same as in Fig. 5.12. The lines inserted in (c) show the energetic positions and oscillator strengths (in arbitrary units) of the one-photon (dashed) and two-photon transitions (solid). From Hu (1991).

To explain this feature, let us recall that in the strong confinement approximation (Sec. 3-1), the first two transition peaks in the one-photon spectrum correspond to the creation of electron-hole pairs in the states

$(1s_e, 1s_h)$, and $(1p_e, 1p_h)$, see Fig. 5.11. Correspondingly, the first two transition peaks in the two-photon spectrum are due to the electron-hole-pair states $(1s_e, 1p_h)$ and $(1p_e, 1s_h)$, respectively. The energy differences between the one- and two-photon absorption peaks are related to the ratio of electron and hole effective masses by

$$\frac{E_{(1p_h - 1s_e)} - E_{(1s_h - 1s_e)}}{E_{(1s_h - 1p_e)} - E_{(1s_h - 1s_e)}} = \frac{E_{(1p_h - 1p_e)} - E_{(1s_h - 1p_e)}}{E_{(1p_h - 1p_e)} - E_{(1p_h - 1s_e)}} = \frac{m_e}{m_h},$$

(5.62)

where $E_{(n_h - n_e)}$ is the energy of the transition in which an electron-hole pair in the state (n_e, n_h) is created. Using the masses for CdS, the first two-photon resonances are situated between the first two one-photon transitions, as seen in Fig. 5.12.

Since these results are in contradiction with experimental observations [Kang et al. (1992)], the band coupling effects discussed in Sec. 4-5 were included in the analysis of the two-photon transitions. An example of the computed two-photon absorption spectra is plotted in Fig. 5.13 showing that the energetically lowest two-photon peak is very close to the linear absorption peak, in contrast to parabolic band results in Fig. 5.12. The one- and two-photon peaks are nearly degenerate because the first excited state of the hole is very close to its ground state for the coupling strength $\mu \simeq 0.7$, as shown in Fig. 4.19. As a consequence, the difference between the energy of the pair states $(1s_e, 1s_h)$ and $(1s_e, 1p_h)$ is reduced in comparison to the value resulting from the parabolic band approximation. This conclusion explains qualitatively the experimental two-photon spectrum [Kang et al. (1991)] in which no observable shift between the one- and two-photon transition peaks occurs. Using similar experimental techniques Tommasi et al. (1992) studied two-photon transitions also in $CdS_x Se_{1-x}$ doped glasses. They observe a rich spectrum of two-photon resonances.

Chapter 6
OPTICAL PROPERTIES OF LARGE DOTS

In this chapter we discuss the optical nonlinearities in large quantum dots, such as they are realized for CuCl crystallites. CuCl is a model system for large quantum dots because the (Z_3) exciton Bohr radius in the bulk material is very small $(a_B \simeq 10$ Å$)$ and the exciton Rydberg energy is quite large $(E_R \simeq 200$ meV$)$. Therefore, 100 Å microcrystallites are already large in comparison to the exciton radius. Optical pump-probe experiments on CuCl microcrystallites [Gilliot (1989) and (1990)] showed interesting nonlinear properties in the excitonic domain which are significantly different from those of the bulk CuCl material. To analyze these properties, we apply in this chapter the simple weak confinement theory, similar to the discussion of Sec. 3-3.

We have seen in Sec. 3-3 that the theory of one electron-hole-pair states in large quantum dots $(\lambda \equiv a_B/R << 1$ - weak confinement limit) can be developed along the original ideas of Efros and Efros (1982), using the confinement quantization of the electron-hole center of mass coordinate. Consequently, we describe bound electron-hole pairs, whose relative motion is in the $1s$ ground state, as Bosons which are characterized through the quantum numbers ℓ,m,n of the confined center-of-mass-motion wavefunction. Typically, there are thousands of such states within an energy range of one E_R.

For application to the CuCl quantum dot problem, this elementary approach has to be modified to deal also with two or more electron-hole pairs. For this purpose, we have to distinguish between bound and unbound two-electron-hole-pair states. The wavefunction for the relative motion in bound states (molecular states or biexcitons) is described by the bound-state wavefunction of the bulk biexciton which has the relatively large binding energy of $\simeq 33$ meV in bulk CuCl. The relative motion for states of independent excitons is restricted only by the boundaries of the microcrystallites. Both aggregates, exciton and biexciton, may be treated assuming confinement quantization of their center-of-mass (cm) motions.

In the following discussion we concentrate on the spectrum of two unbound excitons, but we try to take into account corrections due to the interaction between them.

6-1. Population Induced Nonlinearities

Products of center-of-mass confined exciton states are good candidates for the states of two unbound excitons in the dot after the proper (anti-) symmetrization of the wavefunctions required by the Pauli principle. Such an approach can be viewed as application of the Heitler-London scheme, which may describe also the molecular states if the energy is minimized through the introduction of an envelope function for the two cm coordinates. To simplify the notations, we denote the ensemble of quantum numbers ℓ, m, n with a single symbol ν and the electron-hole-pair (exciton) wavefunction as

$$\Psi_\nu(\mathbf{x}_e, \mathbf{x}_h) = \phi(\mathbf{x}) \, \psi_\nu(\mathbf{X}) ; \quad (\nu = \{n, \ell, m\}). \tag{6.1}$$

We illustrate the theoretical analysis first by considering the electrons and holes as spinless Fermions, or what is equivalent, assuming them to be in the same spin state. For this case only the antisymmetric wavefunction product has to be considered,

$$\Psi_{\nu_1 \nu_2}(\mathbf{x}_{e_1}, \mathbf{x}_{e_2}, \mathbf{x}_{h_1}, \mathbf{x}_{h_2}) = \frac{1}{S} \left[\Psi_{\nu_1}(\mathbf{x}_{e_1}, \mathbf{x}_{h_1}) \, \Psi_{\nu_2}(\mathbf{x}_{e_2}, \mathbf{x}_{h_2}) \right.$$

$$+ \Psi_{\nu_1}(\mathbf{x}_{e_2}, \mathbf{x}_{h_2}) \, \Psi_{\nu_2}(\mathbf{x}_{e_1}, \mathbf{x}_{h_1}) - \Psi_{\nu_1}(\mathbf{x}_{e_1}, \mathbf{x}_{h_2}) \, \Psi_{\nu_2}(\mathbf{x}_{e_2}, \mathbf{x}_{h_1})$$

$$\left. - \Psi_{\nu_1}(\mathbf{x}_{e_2}, \mathbf{x}_{h_1}) \, \Psi_{\nu_2}(\mathbf{x}_{e_1}, \mathbf{x}_{h_2}) \right], \tag{6.2}$$

where the normalization constant is given by

$$|S|^2 = 4 \int d^3x_{e_1} d^3x_{e_2} d^3x_{h_1} d^3x_{h_2} \; \Psi^*_{\nu_1}(\mathbf{x}_{e_1},\mathbf{x}_{h_1}) \; \Psi^*_{\nu_2}(\mathbf{x}_{e_2},\mathbf{x}_{h_2})$$

$$\times \left[\Psi_{\nu_1}(\mathbf{x}_{e_1},\mathbf{x}_{h_1}) \; \Psi_{\nu_2}(\mathbf{x}_{e_2},\mathbf{x}_{h_2}) + \Psi_{\nu_1}(\mathbf{x}_{e_2},\mathbf{x}_{h_2}) \; \Psi_{\nu_2}(\mathbf{x}_{e_1},\mathbf{x}_{h_1}) \right.$$

$$\left. - \Psi_{\nu_1}(\mathbf{x}_{e_1},\mathbf{x}_{h_2}) \; \Psi_{\nu_2}(\mathbf{x}_{e_2},\mathbf{x}_{h_1}) - \Psi_{\nu_1}(\mathbf{x}_{e_2},\mathbf{x}_{h_1}) \; \Psi_{\nu_2}(\mathbf{x}_{e_1},\mathbf{x}_{h_2}) \right]$$

(6.3)

The wavefunction (6.2) does not exactly satisfy the boundary conditions of the quantum dot and is also not an exact eigenfunction of the two-electron-hole-pair Hamiltonian,

$$\mathscr{H}^{(2)} = - \frac{\hbar^2}{2m_{e_1}} \nabla^2 - \frac{\hbar^2}{2m_{e_2}} \nabla^2 - \frac{\hbar^2}{2m_{h_1}} \nabla^2 - \frac{\hbar^2}{2m_{h_2}} \nabla^2$$

$$- \frac{e^2}{|\mathbf{x}_{e_1} - \mathbf{x}_{h_1}|} - \frac{e^2}{|\mathbf{x}_{e_2} - \mathbf{x}_{h_2}|} - \frac{e^2}{|\mathbf{x}_{e_1} - \mathbf{x}_{h_2}|}$$

$$- \frac{e^2}{|\mathbf{x}_{e_2} - \mathbf{x}_{h_1}|} + \frac{e^2}{|\mathbf{x}_{e_1} - \mathbf{x}_{e_2}|} + \frac{e^2}{|\mathbf{x}_{h_1} - \mathbf{x}_{h_2}|} .$$

(6.4)

Both these requirements are fulfilled only in some asymptotical sense for small λ. We compute the average of the energy with the wavefunctions (6.2) as

$$E_{\nu_1 \nu_2} = \frac{1}{|S|^2} \int d^3x_{e_1} d^3x_{e_2} d^3x_{h_1} d^3x_{h_2} \; \Psi^*_{\nu_1 \nu_2}(\mathbf{x}_{e_1},\mathbf{x}_{e_2},\mathbf{x}_{h_1},\mathbf{x}_{h_2})$$

$$\times \mathscr{H}^{(2)} \; \Psi_{\nu_1 \nu_2}(\mathbf{x}_{e_1},\mathbf{x}_{e_2},\mathbf{x}_{h_1},\mathbf{x}_{h_2}).$$

(6.5)

This can be written as

$$E_{\nu_1 \nu_2} = \epsilon_{\nu_1} + \epsilon_{\nu_2} + \frac{4}{|S|^2} \int d^3 x_{e_1} d^3 x_{e_2} d^3 x_{h_1} d^3 x_{h_2} \; \Psi_{\nu_1}^*(\mathbf{x}_{e_1},\mathbf{x}_{h_1}) \; \Psi_{\nu_2}^*(\mathbf{x}_{e_2},\mathbf{x}_{h_2})$$

$$\times \left(V_{e_1 e_2} + V_{h_1 h_2} - V_{e_1 h_2} - V_{e_2 h_1} \right)$$

$$\times \left[\Psi_{\nu_1}(\mathbf{x}_{e_1},\mathbf{x}_{h_1}) \; \Psi_{\nu_2}(\mathbf{x}_{e_2},\mathbf{x}_{h_2}) + \Psi_{\nu_1}(\mathbf{x}_{e_2},\mathbf{x}_{h_2}) \; \Psi_{\nu_2}(\mathbf{x}_{e_1},\mathbf{x}_{h_1}) \right], \quad (6.6)$$

where

$$V_{\alpha\beta} = \frac{e^2}{|\mathbf{x}_\alpha - \mathbf{x}_\beta|} . \tag{6.7}$$

As discussed in the previous Chap. 5, we need to compute the matrix elements of the polarization operator

$$P = p_{cv} \int d^3 x \; \psi_e(\mathbf{x}) \; \psi_h(\mathbf{x}) + \text{h.c.} \tag{6.8}$$

to obtain the optical response. Here, p_{cv} is the valence-conduction-band dipole matrix element and ψ_e and ψ_h are the annihilation operators of electrons and holes, respectively. Since we have already chosen the one- and two-pair wavefunctions, we may calculate the nonvanishing matrix elements

$$\langle 0|P^{(-)}|\nu \rangle = p_{cv} \int d^3 x \; \Psi_\nu(\mathbf{x}, \mathbf{x}) \tag{6.9}$$

and

$$\langle \nu|P^{(-)}|\nu_1 \nu_2 \rangle = p_{cv} \int d^3 x \, d^3 x_e \, d^3 x_h \; \Psi_\nu^*(\mathbf{x}_e, \mathbf{x}_h) \; \Psi_{\nu_1 \nu_2}(\mathbf{x},\mathbf{x}_e,\mathbf{x},\mathbf{x}_h) , \tag{6.10}$$

where $|0\rangle$ is the ground state vector (electron-hole vacuum) and $P^{(-)}$ is the polarization part associated with the annihilation of an electron-hole pair.

The calculation of the matrix elements cannot be done analytically, and is even numerically a formidable task, involving up to twelve-dimensional integrals. However, we should remember that the Efros and Efros (1982) approximation is valid only asymptotically in the parameter λ (see Sec. 3-3). Therefore, we need to consider only the leading terms in λ of the integrals (6.10). This leading-term expansion yields more compact expressions for the characterization of the two-pair states. One finds for the energies

$$E_{\nu_1 \nu_2} = \epsilon_{\nu_1} + \epsilon_{\nu_2} + E^x_{\nu_1 \nu_2} , \qquad (6.11)$$

where ϵ_ν are the one-pair energies,

$$\epsilon_n = E_g - E_R + \frac{m_e \, m_h}{(m_e + m_h)^2} \left(\frac{n \, \pi \, a_B}{R} \right)^2 E_R . \qquad (6.12)$$

The exchange energy is

$$E^x_{\nu_1 \nu_2} = \frac{54.454}{1 + \delta_{\nu_1 \nu_2}} E_R \, \lambda^3 \int d^3X \, |\chi_{\nu_1}(\mathbf{X}) \, \chi_{\nu_2}(\mathbf{X})|^2 . \qquad (6.13)$$

The direct Coulomb term is exponentially small and was ignored.

The polarization matrix elements are given by

$$\langle 0| \, P^{(-)} \, |\nu\rangle = p_{cv} \, g^d_\nu \qquad (6.14)$$

$$\langle \nu| \, P^{(-)} \, |\nu_1 \nu_2\rangle = \frac{p_{cv}}{\sqrt{1 + \delta_{\nu_1 \nu_2}}} \left(\delta_{\nu \nu_2} \, g^d_{\nu_1} + \delta_{\nu \nu_1} \, g^d_{\nu_2} + g^x_{\nu \nu_1 \nu_2} \right) , \qquad (6.15)$$

where

$$g^d_\nu = \lambda^{-3/2} \, \pi^{-1/2} \int d\mathbf{X} \, \chi_\nu(\mathbf{X}) \qquad (6.16)$$

and

$$g^x_{\nu \nu_1 \nu_2} = - \lambda^{3/2} \, 7\pi^{1/2} \int d^3X \; \chi^*_\nu(\mathbf{X}) \; \chi_{\nu_1}(\mathbf{X}) \; \chi_{\nu_2}(\mathbf{X}) \, . \tag{6.17}$$

In Eqs. (6.13) - (6.17) we take the cm integrals and the normalization of the wavefunctions both inside the unit sphere. The only R-dependence is in the respective λ factors. Since the wavefunctions χ_ν of the cm motion are eigenstates of the angular momentum operator, strong selection rules result from the rotational invariance.

If the particle spin is taken into account additional quantum numbers have to be included. Besides the quantum numbers ν_1, ν_2 a two electron-hole-pair state has to be characterized by the total spin state of the two electrons s^e, s^e_z and holes s^h, s^h_z. However, there is degeneracy with respect to s^e_z and s^h_z. Therefore, the two electron-hole-pair states are

$$|\nu_1 \nu_2 ; s^e, s^e_z; s^h, s^h_z \rangle \, , \tag{6.18}$$

where s^e and s^h can take the values 0 and 1, which are the only possible combinations of two 1/2 spins. However, for $\nu_1 = \nu_2$ it is not possible to have the combination $s^e = 0$, $s^h = 1$ or $s^e = 1$, $s^h = 0$.

Retaining again only the leading terms in the λ expansion one finds the energy eigenvalues as

$$E_{\nu_1 \nu_2 \, s^e \, s^h} = \epsilon_{\nu_1} + \epsilon_{\nu_2} + (s^e + s^h - 1) E^x_{\nu_1 \nu_2} \, . \tag{6.19}$$

The polarization matrix elements are

$$\langle \nu; \sigma_e \, \sigma_h | \, P^{(-)} \, | \nu_1 \nu_2 \, ; s^e, s^e_z; s^h, s^h_z \rangle = \left(\zeta^{s^e \, s^e_z} \, \zeta^{s^h \, s^h_z} \right)_{\sigma_e \, \sigma_h}$$

$$\times \frac{P_{cv}}{\sqrt{1+\delta_{\nu_1 \nu_2}}} \left[\delta_{\nu \nu_2} \, g^d_{\nu_1} + \delta_{\nu \nu_1} \, g^d_{\nu_2} + (s^e + s^h - 1) \, g^x_{\nu \nu_1 \nu_2} \right] \, ,$$

$$\tag{6.20}$$

where the matrices ζ are defined as

$$\zeta^{00} = \begin{pmatrix} 0 & 1/\sqrt{2} \\ -1/\sqrt{2} & 0 \end{pmatrix}, \qquad \zeta^{10} = \begin{pmatrix} 0 & 1/\sqrt{2} \\ 1/\sqrt{2} & 0 \end{pmatrix},$$

$$\tag{6.21}$$

$$\zeta^{11} = \begin{pmatrix} 1 & 0 \\ 0 & 0 \end{pmatrix}, \qquad \zeta^{1-1} = \begin{pmatrix} 0 & 0 \\ 0 & 1 \end{pmatrix}.$$

Hence, we see that most of the eigenstates, i.e. nine of the possible ten for equal ν or of the possible sixteen for unequal ν, correspond to the simplified description without spin.

We could attempt to continue the construction of three- and many-pair wavefunctions along the same scheme, but this would be too cumbersome. Instead, we resort to model the many-pair problem within a simplified Boson picture. Since the correct inclusion of spin in a Bose description of the exciton is always problematic [Belleguie and Bányai (1991)], we restrict this discussion to a spinless model. Such a description is known to be useful for some aspects also in bulk materials [see Haug and Schmitt-Rink (1984) for a review].

We introduce Bose creation and annihilation operators b_ν, $b_\nu^\dagger$ which have the previously discussed cm quantum numbers ν. The Hamiltonian of the Bose system is chosen to have the same one- and two-Boson spectrum as the one- and two-electron-hole-pair states of the asymptotic approximation, Eq. (6.12) and (6.13). The existence of the exchange energy implies then the introduction of a diagonal Boson–Boson interaction energy. Thus, we are lead to the Bose Hamiltonian

$$\mathcal{H} = \sum_\nu \epsilon_\nu \, b_\nu^\dagger \, b_\nu + \frac{1}{2} \sum_{\nu\nu'} E_{\nu\nu'}^x \, b_\nu^\dagger \, b_{\nu'}^\dagger \, b_{\nu'} \, b_\nu \, . \tag{6.22}$$

The energies $E_{\nu_1\nu_2}$ are exact energy eigenvalues of the model.

Similarly, we define the polarization operator in terms of our Boson operators as

$$\mathbf{P}^{(-)} = p_{cv} \left[\sum_\nu g_\nu^d \, b_\nu + \sum_{\nu\nu'\nu''} g_{\nu\nu'\nu''}^x \, b_\nu^\dagger \, b_{\nu'} \, b_{\nu'} \right], \tag{6.23}$$

in order to get the same matrix elements as in Eqs. (6.14) - (6.15).

The characteristic features of this model, as compared to the elementary version described in Sec. 3-3, are the existence of an exchange polarization $g^x_{\nu\nu'\nu''}$ and an exchange interaction $E^x_{\nu\nu'}$ between the Bosons. As consequence of the higher order of the small parameter λ contained in their definition, these exchange terms represent small corrections, unless they are amplified by corresponding powers of occupation numbers giving rise to density-dependent effects. The exchange polarization is nothing but the process of pair annihilation where the electron and hole belong to different excitons. This process is shown in Fig. 6.1. According to Eq. (6.15) it opens a number of new transitions which otherwise would have been forbidden due to angular momentum conservation. Indeed, Eq. (6.14) allows only $\ell = 0$ states, whereas in Eq. (6.15) many angular momentum combinations are possible, even if one of the states is an s-state.

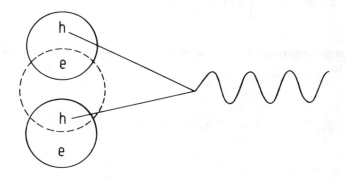

Fig. 6.1: Pair annihilation from different excitons.

One important word of caution in connection with the application of the Bose model relates to the correct treatment of the invariance against rotations. The original quantum mechanical Hamiltonian was rotation invariant, but the classical approximation which is diagonal in the occupation numbers, has a privileged direction. Indeed, due to degeneracy, there is an arbitrariness in the choice of the one-pair basis functions for given ℓ. All orthonormalized linear combination of the $2\ell + 1$ states are equally good. Clearly, the assumption that only the diagonal matrix elements of the Hamiltonian have to be retained cannot hold equally well for all choices of this basis. Assuming that the matrix elements are diagonal within a certain choice implicitly introduces a preferred direction. Only the states in which all the $2\ell + 1$ arbitrary states have equal occupation numbers have energies which are independent of the choice of the basis.

To eliminate the dependence on the choice of the basis, we introduce an average over all different irreducible representations. In a first step we write the classical energy of the system explicitly, without second quantization notation, as

$$E(n) = \sum_{\nu} \epsilon_{\nu} \, n_{\nu} + \frac{1}{2} \sum_{\nu\nu'} I^x_{\nu\nu'} \, n_{\nu} \, n_{\nu'} - \frac{1}{4} \sum_{\nu} I^x_{\nu\nu} \, n_{\nu} \, (n_{\nu} + 1) \, ,$$

(6.24)

where

$$I^x_{\nu\nu'} = 54.454 \, E_R \, \lambda^3 \int d^3X \, |\chi_{\nu_1}(\mathbf{X}) \, \chi_{\nu_2}(\mathbf{X})|^2 \, .$$

(6.25)

For occupation numbers which do not prefer any state m Eq. (6.24) is equivalent to

$$E(n) = \sum_{\nu} \epsilon_{\nu} \, n_{\nu} + \frac{1}{2} \sum_{\nu\nu'} J^x_{\nu\nu'} \, n_{\nu} \, n_{\nu'} - \frac{1}{4} \sum_{\nu} J^x_{\nu} \, n_{\nu} \, (n_{\nu} + 1) \, ,$$

(6.26)

with

$$J^x_{\nu\nu'} = \frac{1}{(2\ell+1)(2\ell'+1)} \sum_{mm'} I^x_{\nu\nu'}$$

(6.27)

and

$$J^x_{\nu} = \frac{1}{2\ell+1} \sum_{m} I^x_{\nu\nu} \, .$$

(6.28)

In the following analysis we use this basis averaged expression of the coefficients in the energy for all possible configurations.

We are interested to study the modification of the optical absorption spectra in the vicinity of the bulk exciton due to the existence of an induced pair population. This population is created by a strong optical pump-field and is assumed to be in equilibrium. This system is then optically tested by a weak field whose frequency lies in the vicinity of the bulk exciton line.

According to linear response theory, the absorption of the test-field is given by

$$\alpha\,(\omega) \propto \frac{1}{R^3} \sum_{\alpha,\beta} \rho^0_{\alpha\beta} \left[|P^{(-)}_{\alpha\beta}|^2\, \delta(E_\alpha - E_\beta - \hbar\omega) - |P^{(-)}_{\beta\alpha}|^2\, \delta(E_\alpha - E_\beta + \hbar\omega) \right].$$

(6.29)

Here the indices α and β run over all the many-body eigenstates of the Hamiltonian $\mathscr{H}$ and ρ^0 stands for the equilibrium density matrix. The negative term represents induced emission. Homogeneous widths can be introduced by smearing out the energy conserving delta functions.

If ρ^0 describes the vacuum state, the exchange terms of the model are irrelevant and one gets the unperturbed or linear absorption corresponding to the creation of a single Boson from the vacuum in an s-state with an arbitrary radial quantum number n,

$$\alpha_{lin}\,(\omega) \propto \frac{1}{R^3} \sum_{n} |g^d_{0,0,n}|^2\, \delta(\epsilon_{0,n} - \hbar\omega)\,.$$

(6.30)

If ρ^0 describes a state with a finite number of excitons in the sphere, many-body effects become important.

Fig. 6.2 shows absorption curves for mixed states with zero or one electron-hole pair for increasing weight of the one-pair state in a microsphere of average radius $R = 10\ a_B$. As long as we restrict ourselves to states with at most one pair, we may also include the spin. Since the energies are expressed in units of the exciton Rydberg energy E_R, the results depend only on the effective mass ratio of the electrons and holes and not on the effective masses themselves. In these calculations the value $m_e/m_h = 0.15$ appropriate for CuCl was chosen, but the results are not very sensitive to this number. The distribution of the one-pair states is taken to be canonical at the temperature $T = 10$ K. The spectra were attributed a common homogeneous width of $\Gamma = 0.005\ E_R$ and were averaged over the Lifshitz-Slezov size distribution, discussed in the Appendix

(Lifshitz and Slezov (1959)]. However, the inhomogeneous width is almost negligible in the present situation. One may see that, apart from the appearance of a small maximum on the low energy wing due to the $s^e = s^h = 0$ state, the main absorption exhibits a blue shift, saturation and increase of the high energy wing.

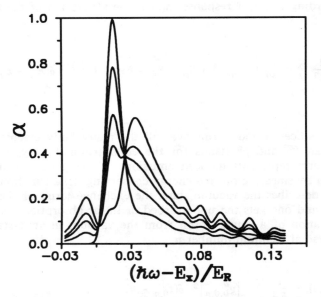

Fig. 6.2 Calculated non-linear absorption with increasing occupation of one-exciton states (spin is taken into account). The fractions of the occupied one-exciton states is 0, 1/3, 2/3 and 1 respectively. From Belleguie and Bányai (1991).

The computed features are also present in the experimental spectra shown in Fig. 6.3. These spectra were obtained [Gilliot *et al.* (1989)] for ≃ 120 Å CuCl microcrystallites in glass by pumping with nanosecond pulses of up to 5 *MW/cm²*. Similar spectra have been obtained by Zimin *et al.* (1990a) (1990b) for the same material under similar conditions except for higher pump intensities. One-beam nonlinear absorption experiments on large CuCl microcrystallites embedded in a NaCl host crystal [Masumoto *et al.* (1989)] and in a glass matrix [Henneberger *et al.* (1988)] have also shown similar blue shift and saturation of the exciton line. The interpretation of one-beam experiments is, however, more difficult since the same light intensity at different frequencies excites different numbers of electron-hole pairs.

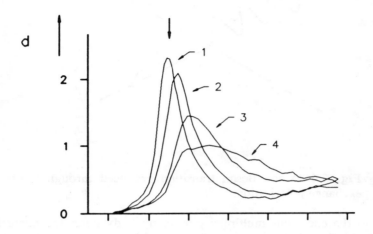

Fig. 6.3 Experimental two-beam absorption in CuCl microcrystallites having a radius $R \simeq 120$ Å in glass by increasing pump-field intensity [Gilliot *et al.* (1989)]. The pump frequency is indicated by the arrow.

For higher pump fields a larger number of excitons is generated. To describe this situation one might consider for ρ_0 the thermal equilibrium distribution

$$\rho^0 = \frac{\exp[-\beta(\mathscr{H} - \mu N)]}{Z} \tag{6.31}$$

where the chemical potential μ and the inverse temperature β are determined by the pump field. Then Eq. (6.29) may be expressed in terms of the many Boson states of our model.

According to the structure of the polarization operator Eq. (6.23) one distinguishes essentially two kind of transitions : i) one-Boson transitions, in which only one occupation number changes its value; ii) many-Boson transitions, in which two or three occupation numbers change their values. They are illustrated in Fig. 6.4.

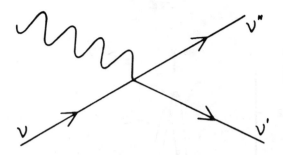

Fig. 6.4 Many Boson transition produced through exchange polarization.

Due to the enormous multiplicity of many-body states the numerical evaluation of the absorption is still impossible without supplementary approximations. We therefore apply a mean field approximation to the expression of the many-body absorption in the sense

$$\sum_{\{n\}} \rho^0{}_{\{n\}} F(\{n\}) \simeq F(\{\langle n \rangle\}) . \tag{6.32}$$

Here, F stands for an arbitrary function of the ensemble of the occupation numbers $\{n\}$ and the average occupation numbers, and $\langle n \rangle$ is given by the equilibrium mean field (Hartree) theory. Within this approximation one has

$$\alpha(\omega) \propto \sum_{nl} F^{(1)}_{nl}(\{\langle n \rangle\}) \, \delta(\mathcal{E}_{nl} - \hbar\omega)$$

$$+ \sum_{nl, n'l', n''l''} F^{(2)}_{nl, n'l', n''l''}(\{\langle n \rangle\}) \, \delta(\mathcal{E}_{nl} + \mathcal{E}_{n'l'} - \mathcal{E}_{n''l''} - \hbar\omega). \tag{6.33}$$

The lengthy explicit expressions of the two kinds of oscillator strengths will not be given here. It is important however to note that $F^{(1)}$ starts with terms of order λ^0 while $F^{(2)}$ starts with λ^6, but always such λ^6 prefactors are multiplied with squares of occupation numbers and therefore give rise to population density factors.

The equilibrium mean field occupation numbers are determined as self-consistent solutions of the equation

$$\langle n_\nu \rangle = \frac{1}{\exp[\beta(\epsilon_\nu^{MF} - \mu)] + 1}, \tag{6.34}$$

where the mean-field energies are determined from

$$\epsilon_\nu^{MF} = \epsilon_\nu + \sum_{\nu'} J_{\nu\nu'}^x \langle n_{\nu'} \rangle - \frac{1}{2} J_\nu^x \left[\langle n_\nu \rangle + \frac{1}{2} \right], \tag{6.35}$$

[Belleguie and Bányai (1991)]. It is worth mentioning that in the finite Bose system, which is equivalent to a classical lattice gas, the mean field contractions have to be applied as $n_\nu \, n_{\nu'} \to n_\nu \langle n_{\nu'} \rangle + n_{\nu'} \langle n_\nu \rangle$ even for $\nu = \nu'$, in contrast to the formal Wick's recipe which would suggest a factor of 2 for $\nu = \nu'$.

After all these approximations it is possible to perform numerical calculations for different temperatures and chemical potentials, taking into account about 200 ℓ, n levels. Resulting absorption curves in a microsphere with $R = 10 \, a_B$, corresponding to an average number of Bosons in the initial state $\langle N \rangle = 0$, 5 and 10 are given in Fig. 6.5 for the respective temperatures of $T = 0$, 100 and 200 K and homogeneous widths $\Gamma = 0.03$, 0.045 and 0.09 E_R respectively. The temperatures and homogeneous widths are chosen according to the plausible scenario, that with an increase of the pump intensity not only the number of excited Bosons increases, but also their temperature and consequently their energetic broadening. We see that the general features are in a good agreement with the experimental pump-probe results [Gilliot et al. (1989)] of Fig. 6.3.

From the comparison between the theoretical and experimental blue shifts of the absorption curves one may conclude, that the number of pump-induced excitons in these experiments exceeds one but is smaller than five. In the range between these two average numbers, however we have no possibilities to give good evaluations, because mean field works only for $\langle N \rangle \gg 1$. An interesting feature of the theoretical curves of Fig. 6.5 is the existence of a high energy plateau, increasing with the exciton density, due to the opening of a large number of new many-body channels accessible through the exchange polarization described in Fig. 6.4.

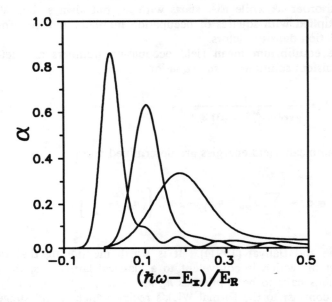

Fig. 6.5 Calculated nonlinear absorption by simultaneous increase of the number of Bosons without spin, of the temperature, and of the homogeneous width. The three curves correspond to $\langle N \rangle = 0$, $T = 0\ K$, $\Gamma = 0.03\ E_R$; $\langle N \rangle = 5$, $T = 100\ K$, $\Gamma = 0.045\ E_R$, $\langle N \rangle = 10$, $T = 200\ K$, $\Gamma = 0.09\ E_R$ respectively.

The weak confinement model applied in this chapter starts from the assumption that the induced pair population is in excitonic bound states. This seems to be justified because of the large exciton binding energy in CuCl. Another kind of problem may arise in the presence of an electron-hole plasma. In bulk semiconductors (including CuCl) one observes experimentally a bleaching of the exciton line with increasing plasma density, but no shift of the exciton. This behavior is explained through the fact that the lowering of the exciton binding energy due to the plasma screening is compensated by the simultaneous lowering of the bound gap (negative variation of the Coulomb self-energy). Nevertheless, the oscillator strength of the exciton is reduced and at the Mott density the exciton bound state disappears. A bump survives for a while, due to the still strong Coulomb correlation (Coulomb enhancement).

It is interesting to see whether this scenario still holds in large quantum dots. Since screening weakens the exciton binding energy giving rise to enlarged effective exciton radii, it is expected [Bányai and Koch (1986)] that a screened exciton is influenced more strongly by the boundaries of the dot. Let us write the effective exciton energy in the dot in the presence of an electron-hole plasma as

$$E_x(n,R) = E_g(n,R) + \epsilon(n,R) \tag{6.36}$$

where $E_g(n,R)$ is the effective (renormalized) band gap and $\epsilon(n,R)$ the exciton binding energy. The density dependent exciton binding energy is the eigenvalue of the Wannier equation with screening and boundary conditions or more generally of the Bethe-Salpeter equation including besides screening also final state occupation effects [Haug and Schmitt-Rink (1984)]. Both energies depend on the electron-hole plasma density n and the dot radius R. We are interested in the variation

$$\delta E_x = E_x(n,R) - E_x(0,R) . \tag{6.37}$$

On the other hand we know that this shift vanishes for $R \to \infty$. Now, if one assumes [Bányai and Koch (1986)], that the band gap variation in large dots is already independent of the radius, then

$$\delta E_x \simeq \epsilon(n,R) - \epsilon(n,\infty) - \epsilon(0,R) + \epsilon(0,\infty) . \tag{6.38}$$

The last two terms are the confinement energy at the given radius in the absence of screening, whereas the first two terms are the confinement energy in the presence of screening. The confinement energy varies between the confinement energy of the free electron - hole pair (for $R/a_B \ll 1$)

$$\left(\frac{\pi a_B}{R}\right)^2 E_R$$

and the confinement energy of the exciton (for $R/a_B \gg 1$)

$$\frac{m_e \, m_h}{(m_e + m_h)^2} \left(\frac{\pi a_B}{R}\right)^2 E_R .$$

Now, it is to be expected, that for a radius $R > a_B$ with strong screening one has free particle confinement, while without screening only the weaker exciton confinement holds. Therefore, we get the upper bound

$$\delta E_x \leq \left[1 - \frac{m_e \, m_h}{(m_e + m_h)^2}\right] \left(\frac{\pi a_B}{R}\right)^2 E_R \,. \tag{6.39}$$

Our numerical estimates are close to this upper bound.

We should mention here, that the experimental results for band to band excitation in large CuCl quantum dots seem to be somewhat contradictory. While for dots embedded in a glass matrix in two-beam experiments no nonlinearity at all was observed for pump intensities up to close to the damage threshold [Zimin et al. (1990b)], dots with a radius of 62 Å embedded in NaCl show exciton bleaching and a blue shift of about 0.03 E_R around the total bleaching [Masumoto et al. (1990)], which is at least compatible with Eq. (6.39) which yields for this case a shift of 0.1 E_R. On the other hand, even in a NaCl matrix one sees no increase of the absorption at the high energy side of the exciton, which would indicate a red shift of the band typical for plasma screening. The differences in the two quoted experiments may be attributed to the fact that in glass the surrounding barrier may be lower and partially transparent for the electrons and holes so that the optically generated electron-hole pairs diffuse into the glass, but this is only speculation at this point. Clearly, more experiments and detailed material characterizations are needed.

6-2. Third-Order Nonlinearities

In this section we discuss the third-order susceptibility of large quantum dots, whose radius R substantially exceeds the exciton Bohr radius a_B of the bulk semiconductor [Belleguie and Bányai (1993)]. The typical feature of this case is the small energy separation between allowed states which requires that a large number of states has to be taken into account, making the direct numerical scheme of the previous chapter very difficult. For the large quantum dots some sort of analytical approximation is unavoidable.

We discussed in Sec. 3-3 that the typical approximation for large quantum dots is to assume quantization of the exciton cm motion [Efros and Efros (1982)]. The straightforward extension of this method to many-pair states is to assume that the electron-hole pairs are ideal non-interacting Bosons, see previous Sec. 6-1. If we furthermore take all decay parameters as equal,

$$\Gamma^0{}_i = \gamma \ , \tag{6.40}$$

then we have a very simple model, which however has a vanishing third order susceptibility [Hanamura (1987)]. Moreover, it is well-known, that such a model has no optical nonlinearities at all [Kranz and Haug (1986); Bányai *et al.* (1988)].

A deviation from this simple exciton-Boson picture is needed in order to obtain a finite third-order susceptibility. For example, the introduction of the molecular bound state (bound biexciton state) gives rise to the typical biexcitonic nonlinearities. The corresponding polarization matrix elements have no volume enhancement factors and therefore their discussion is relatively straightforward. Biexcitonic nonlinearities are not expected to give rise to strong volume dependent effects.

In the following discussion we analyze in more detail the properties of the slightly improved Boson model without spin in which exchange interactions between the excitons as well as exchange polarizations are taken into account. We furthermore include the possibility that the decay parameters Γ_i are not necessarily all identical. Simple algebra shows, that the inclusion of the exchange interaction together with state-dependent decay parameters gives rise to the nonlinear susceptibility

$$\chi_3^{sym}(\omega_P, -\omega_P, \omega_T)_1 =$$

$$\frac{1}{8\pi a_B^3} \sum_{e,e'} (g_e^d \, g_{e'}^d)^2 \, [\, V_{e,e'} + i \, \lambda^{-3} \, (\Gamma_{o,e} - \Gamma_{(e,e'),e'}) \,] \, A_{e,e'} \tag{6.41}$$

where

$$A_{e,e'} = \frac{1}{(\epsilon_e - \hbar\omega_T - i \, \Gamma_{oe}) \, (\epsilon_e + V_{e,e'} - \hbar\omega_T - i \, \Gamma_{(e,e'),e})} \, (F_{e,e'} + G_{e,e'}) \ ; \tag{6.42}$$

$$F_{e,e'} = \frac{2\Gamma_{e'o}/\Gamma_{e'e'}}{\Gamma_{e'o}^2 + (\epsilon_{e'} - \hbar\omega_P)^2} + \left[\frac{1}{\Gamma_{e'o} - i(\epsilon_{e'} - \hbar\omega_P)} + \frac{1}{\Gamma_{e'o} + i(\epsilon_{e'} - \hbar\omega_T)} \right]$$

$$\times \frac{1}{\Gamma_{e'e'} - i(\hbar\omega_T - \hbar\omega_P)} + \frac{1}{\Gamma_{e'o} - i(\epsilon_{e'} - \hbar\omega_P)} \left(\frac{1}{\Gamma_{ee'} + i(\epsilon_e + \epsilon_{e'} + V_{e,e'})} \right.$$

$$+ \frac{1}{\Gamma_{ee'} + i((\epsilon_e + \epsilon_{e'} + V_{e,e'}) - \hbar\omega_T + \hbar\omega_P)} \bigg) \tag{6.43}$$

and

$$G_{e,e'} = \frac{1}{\epsilon_e + \epsilon_{e'} + V_{e,e'} - \hbar\omega_T - \hbar\omega_P - i\,\Gamma_{(e,e'),e'}} \left(\frac{1}{\epsilon_e - \hbar\omega_T - i\,\Gamma_{oe}} \right.$$

$$+ \frac{1}{\epsilon_e - \hbar\omega_P - i\,\Gamma_{oe}} - \frac{1}{\epsilon_{e'} - \hbar\omega_T - i\,\Gamma_{oe'}} - \left. \frac{1}{\epsilon_{e'} - \hbar\omega_P - i\,\Gamma_{oe'}} \right). \qquad (6.44)$$

Here, all the explicit dependencies on the dot radius were converted into powers of $\lambda = a_B/R$, with the exception of the radius dependence of the one-Boson energies ϵ_e in the expression of $A_{e,e'}$.

An open question is the radius dependence of the decay parameters Γ. Inspecting Eq. (6.41) we can convince ourselves that either the difference of the one-exciton and two-exciton polarization decay rates vanishes at least as fast as the inverse volume of the dot

$$\Gamma_{o,e} - \Gamma_{(e,e'),e'} \propto \lambda^3 \quad,$$

or the susceptibility is not properly defined, i.e., it will be extensive (increasing with the volume). Since our model is not a pure Hamiltonian model such difficulties are not unexpected. For the consistency of the model we therefore have to assume that

$$\Gamma_{o,e} - \Gamma_{(e,e'),e'} = \lambda^3\,\gamma_{e,e'} \quad, \qquad (6.45)$$

where $\gamma_{e,e'}$ remains finite when $\lambda \to 0$. Then

$$\chi_3^{sym}(\omega_P, -\omega_P, \omega_T)_1 = \frac{1}{8\pi a_B^3} \sum_{e,e'} (g_e{}^d\,g_{e'}{}^d)^2\,[\,V_{e,e'} + i\,\gamma_{e,e'}\,]\,A_{e,e'}. \qquad (6.46)$$

Note, that even without the explicit exciton-exciton interaction energy $V_{e,e'}$ one would obtain a finite third-order susceptibility due to the term $\gamma_{e,e'}$, which is actually an implicit consequence of the exciton-exciton interaction.

The inclusion of the exchange polarization δP gives rise to another nonlinearity. Since δP is already of order $\lambda^{3/2}$, the inclusion of a single δP correction gives an intensive contribution. The inclusion of a further second-order δP correction, however may be omitted. Then we can write the second piece of the third-order susceptibility as

$$\chi_3^{sym}(\omega_P,-\omega_P,\omega_T)_2 = \frac{1}{4\pi a_B^3} \sum_{e,e',e''} g_e^{\ d}\ g_{e'}^{\ d}\ g_{e''}^{\ d}\ g_{e,e',e''}^x\ B(e,e',e'')$$

(6.47)

where

$$B(e,e',e'') = C_{(e,e'),e''} + C_{(e,e'),e''}$$ (6.48)

and

$$C_{(e,e'),e''} = \frac{1}{\Gamma_{(e,e')o}+i(\epsilon_e+\epsilon_{e'}+V_{e,e'}-\hbar\omega_P-\hbar\omega_T)}$$

$$\times \left(\frac{1}{\Gamma_{eo}+i(\epsilon_e-\hbar\omega_T)} - \frac{1}{\Gamma_{(e,e')e''}+i(\epsilon_e+\epsilon_{e'}+V_{e,e'}-\epsilon_{e''}-\hbar\omega_T)}\right)$$

$$\times \left(\frac{1}{\Gamma_{e''o}+i(\epsilon_{e''}-\hbar\omega_P)} + \frac{1}{\Gamma_{e''o}+i(\epsilon_{e''}-\hbar\omega_T)}\right)$$

$$- \frac{1}{\Gamma_{(e,e')e}+i(\epsilon_{e'}+V_{ee'}-\hbar\omega_T)}\left[\frac{1}{\Gamma_{eo}-i(\epsilon_e-\hbar\omega_P)}\right.$$

$$\times \left(\frac{1}{\Gamma_{ee''}-i(\epsilon_e-\epsilon_{e''})} + \frac{1}{\Gamma_{ee''}-i(\epsilon_e-\epsilon_{e''}+\hbar\omega_T-\hbar\omega_P)}\right)$$

$$+ \frac{1}{\Gamma_{e''o}+i(\epsilon_{e''}-\hbar\omega_P)}\ \frac{1}{\Gamma_{ee''}-i(\epsilon_e-\epsilon_{e''})}$$

$$+ \left.\frac{1}{\Gamma_{e''o}+i(\epsilon_{e''}-\hbar\omega_T)}\ \frac{1}{\Gamma_{ee''}-i(\epsilon_e-\epsilon_{e''}+\hbar\omega_T-\hbar\omega_P)}\right].$$ (6.49)

Both pieces of the third order susceptibility are smooth functions of the dot radius. The enhancement factors appearing in the unperturbed polarization have been completely compensated. Whether a significant enhancement of the third-order susceptibility with increasing dot radius occurs can be decided only by analyzing the radius dependence of Eqs. (6.46) - (6.47). The critical point of the theory is the fact that the predicted nonlinearity may strongly depend on the knowledge of a very small correction to the decay parameters which vanishes in the bulk material. We will illustrate this point later.

After the introduction of the explicit expressions of the relevant matrix elements, the two contributions to the third-order susceptibility are given as

$$\chi_3^{sym}(\omega_P, -\omega_P, \omega_T)_1 = -\frac{P_{cv}^4}{(a_B E_R)^3} \frac{52}{3\pi^3} \int_0^1 dr \frac{1}{r^2}$$

$$\times \sum_{n_1 = 1}^{\infty} \sum_{n_2 = 1}^{\infty} [\sin(n_1 \pi r) \sin(n_2 \pi r)]^2 A\left(\frac{n_1}{R}, \frac{n_2}{R}; \omega_P, \omega_T\right) \qquad (6.50)$$

and

$$\chi_3^{sym}(\omega_P, -\omega_P, \omega_T)_2 = -\frac{P_{cv}^4}{(a_B E_R)^3} \frac{28}{\pi^4} \int_0^1 \frac{dr}{r} \sum_{n_1 = 1}^{\infty} \sum_{n_2 = 1}^{\infty} \sum_{n_3 = 1}^{\infty} (-1)^{n_1 + n_2 + n_3}$$

$$\times \sin(n_1 \pi r) \sin(n_2 \pi r) \sin(n_2 \pi r) B\left(\frac{n_1}{R}, \frac{n_2}{R}, \frac{n_3}{R}; \omega_P, \omega_T\right). \qquad (6.51)$$

We took here $\gamma_{n_1 n_2} = 0$ and for simplicity we ignored the energy corrections of order λ^3 which occur due to the exciton-exciton interaction.

The quantized exciton energies are

$$\epsilon_n = E_g - E_R + \frac{m_e m_h}{(m_e + m_h)^2} \left[\frac{n \pi a_B}{R}\right]^2 E_R \qquad (6.52)$$

for the $\ell = 0$ states, which are the only ones that contribute. Since the functions A and B in Eqs. (6.50) - (6.51) consist of products of energy denominators (where the energies are measured in units of E_R), they are asymptotically decreasing with n for a fixed radius and become finite functions of ω_P and ω_T as R goes to infinity at fixed n. For constant A and B the expressions (6.50) - (6.51) are well-known, convergent and explicitly summable series. Thus we obtain the asymptotic expression of the susceptibility for $R \to \infty$ as

$$\lim_{R \to \infty} \chi_3^{sym}(\omega_P, -\omega_P, \omega_T) =$$

$$\frac{p_{cv}^4}{(a_B E_R)^3} \left[-3.816\, A(0,0;\omega_P,\omega_T) + 0.3708\, B(0,0,0;\omega_P,\omega_T) \right]. \qquad (6.53)$$

These analytic estimates show that no unbounded enhancement of the third-order susceptibility occurs for increasing radius of the dot. Nevertheless numerical computations [Belleguie and Bányai (1993)] show, that the maxima of susceptibility are very high in comparison to their values at $R/a_B = 10$ and the limiting behavior is achieved only slowly for sizes above $R/a_B = 100$.

Leaving aside the uncertainty contained in the unknown radius dependence of the damping parameters, one may ask whether the asymptotic expression (6.53) can be taken seriously. Indeed, a correct treatment of very large microcrystallites must take into account the existence of the dissociated pair states, as well as the finiteness of the wavelength of the electromagnetic field. This last aspect will be illustrated for the example of the direct recombination of an exciton in the microcrystallites.

In bulk semiconductors, the direct recombination of an exciton without simultaneous phonon emission is forbidden by the conservation of energy and momentum (with the exception of certain momenta, where the exciton-photon mixing occurs) and therefore it occurs mainly through surface or impurity states. In a quantum dot, however, also the direct radiative recombination is allowed. Writing the dipole interaction Hamiltonian in the form of Eq. (6.40) we assumed from the beginning that the wavelength of the photon is negligible in comparison to the radius of the dot. Therefore this form of the dipole Hamiltonian is not suitable for the discussion of radiative decay in the limit of large radii exceeding the wavelength of the emitted radiation. A correct treatment must take into account the local nature of the electromagnetic interaction and the finite wavelength of the photon. Therefore, one should consider the interaction of the electrons and holes in the quantum dot with the radiation field according to the Hamiltonian

$$\mathcal{H}_{int} = p_{cv} \int d^3r\, \psi_e(\mathbf{r})\, \psi_h(\mathbf{r})\, E^{(+)}(\mathbf{r}) + \text{h.c.}, \qquad (6.54)$$

where we used again the annihilation operators for the electrons and holes. The creation part of the electric field operator is

$$E^{(+)}(\mathbf{r}) = \sum_{\mathbf{k}} \sqrt{\frac{\hbar\omega_{\mathbf{k}}}{2V}}\, a_{\mathbf{k}}^{\dagger}\, e^{i\mathbf{k}\cdot\mathbf{r}} \, . \tag{6.55}$$

The volume V here is the formal quantization volume of the photons and has no relation to the volume of the quantum dot. The classical counterpart of the Hamiltonian (6.54) for $k \rightarrow 0$ coincides with the dipole Hamiltonian of Sec. 5-1.

Applying the Golden Rule of quantum mechanics, one may calculate the lifetime of the one-pair state e due to direct radiative recombination in the dot as

$$\Gamma_e = \frac{2\pi}{\hbar} \sum_{\mathbf{k}} |\, \langle 0 |\, \mathscr{H}_{int}\, | e \rangle\, |^2\, \delta(\epsilon_e - \hbar\omega_{\mathbf{k}}) \, . \tag{6.56}$$

Using the energies and eigenstates of the quantum dot, we obtain the result

$$\Gamma_e = \frac{32\pi^2 p_{cv}^2}{\hbar a_B^3}\, \kappa_{n,\ell}^2\, \frac{(kR)^3}{(\kappa_{n,\ell}^2 - k^2 R^2)^2}\, j_\ell(kR)^2 \tag{6.57}$$

where the explicit weak confinement quantum numbers ℓ, m, n of the state e were introduced, j_ℓ are Bessel functions, $\kappa_{n,l}$ are their zeroesy and $k = \epsilon_{n,\ell}/\hbar c$.

For $R \rightarrow \infty$ the argument kR goes to infinity and Γ_e vanishes faster, but not monotonously, as any power of R^{-1} for any finite ℓ and n. This is in perfect agreement with the result for unconfined excitons, which of course may be obtained also with plane waves for the center of mass motion of the exciton. On the other hand, if the photon wavelength is still much bigger than the radius of the sphere, i.e. $kR \ll 1$, we have

$$\Gamma_e \simeq \delta_{\ell,0}\, \frac{32\pi^2 p_{cv}^2}{\hbar a_B^3}\, \frac{(kR)^3}{(n\pi)^2} \, . \tag{6.58}$$

However we should not conclude that the decay rate is enhanced with the radius, since kR was already assumed to be very small. This example shows some of the difficulties associated with the volume dependence of the decay parameters, even within the simplest possible mechanism for the simplest population decay rate.

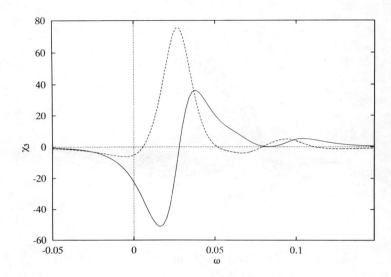

Fig. 6.6: The imaginary (full curve)   and real (dotted   curve)   parts   of $\chi_3^{sym}(w,w,-w)$ 10^{-5} $(a_B E_R)^3/p_{cv}^4$ as function of $w \equiv (\hbar\omega - E_x)/E_R$ at $R/a_B = 10$.

At the end of this chapter we now show some numerical evaluations of the third-order susceptibility [Belleguie and Bányai (1993)]. The susceptibilities are all normalized and therefore dimensionless. Except for the last figure, where a fit to the experiment was attempted, we always use a common broadening parameter $\Gamma = 0.03 E_R$ and a hole-electron mass ratio $m_h/m_e = 4$.

The typical behavior of the single-beam third-order susceptibility in the exciton region is illustrated in Fig. 6.6. The numerical curves (for the real and imaginary parts) are obtained for a dot of radius $R = 10 \, a_B$. The simultaneous dependence on the frequency and the radius of the dot is shown in the three-dimensional plot of Fig. 6.7. We see a slight shift to the origin (E_x) and an enhancement with increasing radius. This enhancement reaches its asymptotic value of about six time the susceptibility at $R = 10 \, a_B$ only for $R \gg 100 \, a_B$ (not represented in the figure). The contribution of the exchange part of the susceptibility was found to be negligible, whereas the energy corrections of order λ^3 in the denominators of the

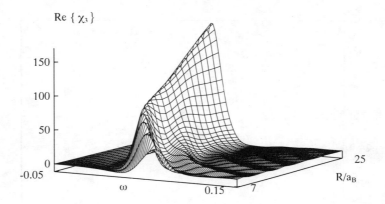

Fig. 6.7a: The real part of $\chi_3^{sym}(w,w,-w)$ 10^{-5} $(a_B E_R)^3/p_{cv}^4$ as function of $w \equiv (\hbar\omega - E_x)/E_R$ and R/a_B.

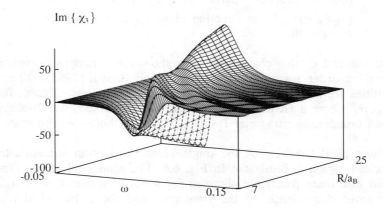

Fig. 6.7b: The imaginary part of $\chi_3^{sym}(w,w,-w)$ corresponding to Fig. 6.7a.

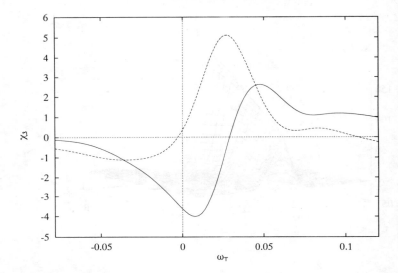

Fig. 6.8a: The imaginary (full curve) and real (dotted curve) parts of $\chi_3^{sym}(w_P, w_T, -w_P) \, 10^{-5}$ $(a_B E_R)^3/p_{cv}^4$ as function of $w_T \equiv (\hbar\omega_T - E_x)/E_R$ at $R/a_B = 10$ for the pump frequency $w_P = -0.05$.

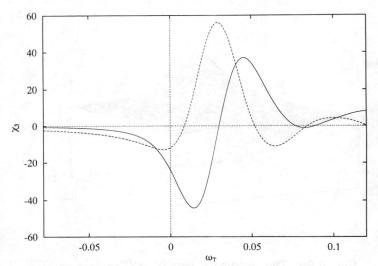

Fig. 6.8b: Same as Fig. 6.8a but for the pump frequency $w_P = 0.01$.

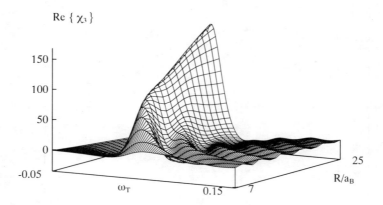

Fig. 6.9a: The real part of $\chi_3^{sym}(w_{P,w_T},-w_P)$ 10^{-5} $(a_B E_R)^3/p_{cv}^4$ as function of $w_T \equiv (\hbar\omega_T - E_x)/E_R$ and R/a_B at fixed $w_P = +0.01$.

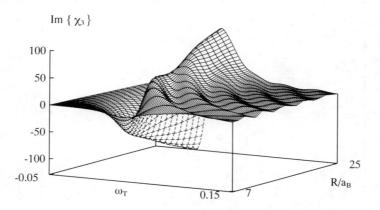

Fig. 6.9b: The imaginary part of $\chi_3^{sym}(w_{P,w_T},-w_P)$ corresponding to Fig. 6.9a.

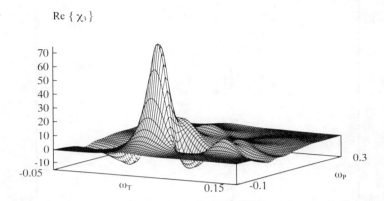

Fig. 6.10a: The real part of $\chi_3^{sym}(w_{P,w_T},-w_P)$ 10^{-5} $(a_B E_R)^3/p_{cv}^4$ as function of $w_T \equiv (\hbar\omega_T - E_x)/E_R$ and $w_P \equiv (\hbar\omega_P - E_x)/E_R$ at fixed $R/a_B = 10$.

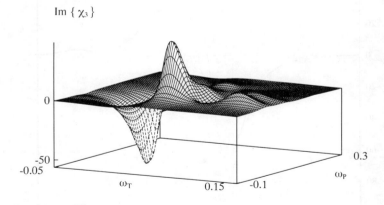

Fig. 6.10b: The imaginary part of χ_3^{sym} corresponding to Fig. 6.10a.

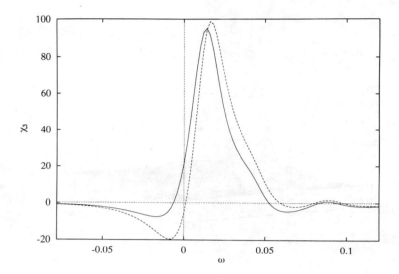

Fig. 6.11a: The real part of $\chi_3^{sym}(w,w,-w)$ 10^{-5} $(a_B E_R)^3/p_{cv}^4$ as function of $w \equiv (\hbar\omega - E_x)/E_R$ at $R/a_B = 15$ with $\gamma_{e,e'} = 0$ (full curves), $\gamma_{e,e'} = 0.5$ $V_{100,100}$ (dashed curves).

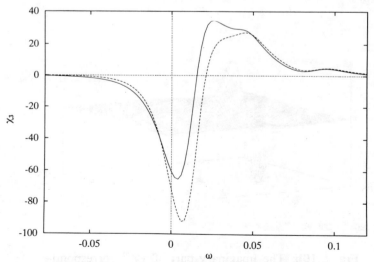

Fig. 6.11b: The imaginary part of $\chi_3^{sym}(w,w,-w)$ corresponding to Fig. 11a.

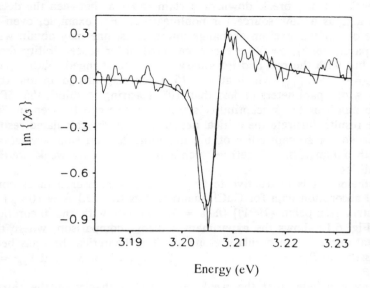

Fig.   6.12:   Experimentally   measured   Im $\chi_3^{sym}(w_{P,w_T},-w_P)$ (arbitrary units) in CuCl and theoretical fit.

direct susceptibility were found to be important for the frequency dependence at moderate radii $R \simeq 10\ a_B$.

The pump-probe third-order susceptibility as a function of the probe frequency for a fixed pump frequency is shown in Fig. 6.8. The enhancement effect with increasing radius in the three-dimensional representation (simultaneous probe-frequency and radius dependence at fixed pump frequency $w_P = 0.01$) is shown in Fig. 6.9. The asymptotic enhancement factor is found to be around eight.

The simultaneous pump- and probe-frequency dependence at fixed $R = 10\ a_B$ is given in Fig. 6.10. The enhancement of the third order susceptibilities with increasing dot radius is due to the accumulation of the poles at the bulk exciton energy and, in contrast to an enhancement through a volume factor, is limited to an overall factor that is less than an order of magnitude between $R = 10\ a_B$ and $R = \infty$. Of course the comments of the previous Section regarding the validity of this theory for very big radii have to be kept in mind.

If we assume a very small difference between the width parameters according to Eq. (6.46), we obtain essential quantitative changes. As we already mentioned, the break-down of certain relations between the decay parameters acts as a new source for nonlinearities. For example, even in the absence of exciton-exciton exchange interactions, one may obtain with a suitable parameter $\gamma_{e,e'} = V_{e,e'}$ the same third order susceptibility from Eq. (6.46), but with the real and imaginary parts interchanged. With a parameter $\gamma_{e,e'} = 0.5 \, V_{100,100}$, which at $R = 15 \, a_B$ corresponds to an uncertainty in the width parameters of less than 10% (bearing in mind, that $\delta\Gamma = \lambda^3 \, \gamma$), one modifies the susceptibility significantly as can be seen in Fig. 6.11. These results illustrate the critical dependence of the calculated results for the third-order susceptibility of big quantum dots on small variations of the width parameters. Therefore, such a theory has only weak predictive capabilities.

Nevertheless, it is instructive to fit the recently measured picosecond differential absorption data for CuCl quantum dots ($R = 70 \, \text{Å} \simeq 10 \, a_B$) in a glass matrix [Kippelen (1991)] ($\hbar\omega_P = 3.2053 \, eV$) with the theoretical formula. Fig. 6.12 shows the experiment - theory comparison, where the experimental curve is the oscillating one and the theoretical line has been obtained assuming $\Gamma_{eo} = 4.5 \, meV$, $\Gamma_{be} = 5 \, meV$, $\Gamma_{b0} = 3.5$ and $\Gamma_{ee'} = 2$ meV.

We may conclude, that the weak confinement theory of the third-order susceptibility of semiconductor microcrystallites is compatible with the experimental data, but due to its inherent sensitivity to small volume-dependent variations of the width parameters it has no strong predictive character. Small variations of the relaxation times that are vanishing in the bulk limit act as sources of new independent nonlinearities of the same magnitude as those due to exciton exchange effects.

Chapter 7
PHONONS AND EXTERNAL FIELDS

In this chapter we discuss the influence of the interaction of optically generated electron-hole pairs in quantum dots with the ionic lattice and with external fields. In Sec. 7-1 we outline some of the theoretical background of the electron-phonon interaction in quantum dots. In an early analysis of this problem Schmitt-Rink *et al.* (1987) used the perfect strong confinement theory with identical electron and hole wavefunctions. As a consequence of this elementary confinement approximation, the conclusion was reached that no polar coupling should exists in quantum dots, whereas other short wavelength phonon interactions were predicted to be enhanced by the factor $(a_B/R)^3$. In contrast to these estimations, we know now that the presence of Coulomb interactions, spin-orbit coupling, and a finite barrier leads to different ground-state charge distributions for electrons and holes even in idealized quantum dots. Hence, a local charge density exists in spite of the overall charge neutrality which leads to substantial electron-phonon coupling.

In Sec. 7-2 we deal with the influence of static electric fields on the electron-hole excitations. Both experimentally and theoretically it is well-known that applied d.c. electric fields along the confinement direction of quantum wells give rise to shifts of the absorption exciton peaks due to the confined Stark effect [Bastard *et al.* (1983), Miller *et al.* (1985)]. The basic phenomena were originally analyzed without considering the Coulomb interaction. However, we will show that this interaction leads to modifications of the electric field phenomena in quantum dots.

In Sec. 7-3 we discuss some impressive experimental and theoretical aspects of fabricated quantum dots in a magnetic field. These dots are actually quasi-two-dimensional islands of about 100 nm size containing between 2 and 2000 electrons. We discuss numerical calculations of the spectra in quantum boxes containing up to six electrons which have shown a great complexity of the level structure [Bryant (1987)]. Furthermore, we briefly outline a number of impressive experimental and theoretical studies involving island quantum dots in magnetic fields.

7-1. LO-Phonons in Quantum Dots

We have seen in Chap. 4 that Coulomb interaction, spin–orbit coupling, and finite barrier effects lead to different ground-state charge distributions for electrons and holes. Most of the semiconductor materials used for quantum dots are highly polar, and therefore a local charge density causes a strong Coulomb interaction with the lattice. Currently, there are many direct and indirect experimental evidences for this coupling [Roussignol *et al.* (1989), Uhrig *et al.* (1990), Klein *et al.* (1990), Wang and Herron (1990)].

In order to prepare the background for our discussion of phonons in quantum dots, we first present the classical continuum model of optical phonons in an infinite medium. In the next step we then introduce the necessary quantum mechanical modifications necessary for a quantum dot.

In the classical continuum model one assumes that the lattice polarization in every point is proportional to the local displacement field of the medium

$$\mathbf{P(r)} = \frac{e}{v} \mathbf{u(r)} \,, \tag{7.1}$$

where e is the unit charge and v is the volume of the elementary cell. The corresponding lattice polarization charge density is

$$\rho_{pol}(\mathbf{r}) = - \nabla \cdot \mathbf{P(r)} = - \frac{e}{v} \nabla \cdot \mathbf{u(r)} \,. \tag{7.2}$$

In the presence of an external charge density one has the Poisson equation for the potential V

$$\epsilon_\infty \nabla^2 V(\mathbf{r}) = - 4\pi \left[\rho_{pol}(\mathbf{r}) + \rho_{ext}(\mathbf{r}) \right] \,, \tag{7.3}$$

where ϵ_∞ is the dielectric constant of the background electrons. Note, that our definition of **P** does not include the electronic polarization.

In equilibrium, Eq. (7.3) may be written as

$$\epsilon_0 \nabla^2 V(\mathbf{r}) = -4\pi \rho_{ext}(\mathbf{r}) \,, \tag{7.4}$$

which serves actually for the definition of the total background dielectric constant ϵ_0. The traditional notations ϵ_∞ and ϵ_0 correspond to the high-frequency (fast response) and low frequency (slow-response) dielectric constants.

The Lagrangeian for the continuous displacements is

$$\mathscr{L} = \int d^3r \, \frac{m}{2v} \left[\left(\frac{d\mathbf{u}}{dt}\right)^2 - \omega_0^2 \, \mathbf{u}^2 \right] - \frac{1}{2\epsilon_\infty} \int d^3r \, d^3r' \, \frac{[\nabla \cdot \mathbf{P}(\mathbf{r})] \, [\nabla \cdot \mathbf{P}(\mathbf{r}')]}{|\mathbf{r} - \mathbf{r}'|}$$

$$+ \frac{1}{\epsilon_\infty} \int d^3r \, d^3r' \, [\nabla \cdot \mathbf{P}(\mathbf{r})] \, \frac{\rho_{ext}(\mathbf{r}')}{|\mathbf{r} - \mathbf{r}'|} \, , \qquad (7.5)$$

where the medium is represented by oscillators with a characteristic frequency ω_0 and mass density m/v, to which the Coulomb interaction terms have been added. The transition from the discrete lattice sums to integrals leads to a supplementary local field term in the potential energy due to the exclusion of a small sphere or cube,

$$\frac{4\pi}{3} \int d^3r \, \mathbf{P}(\mathbf{r})^2 \, .$$

However, this correction can be absorbed in the definition of ω_0.

Using the Lagrangeian (7.5) we obtain the Euler-Lagrange equations. These equation assume a simple form if one writes them for the transverse and longitudinal components of the displacement field

$$\mathbf{u}(\mathbf{r}) = \mathbf{u}_\ell(\mathbf{r}) + \mathbf{u}_t(\mathbf{r}) \, , \qquad (7.6)$$

where

$$\mathbf{u}_\ell(\mathbf{r}) = - \nabla \int d^3r' \, \frac{\nabla' \cdot \mathbf{u}(\mathbf{r}')}{4\pi \, |\mathbf{r} - \mathbf{r}'|} \qquad (7.7)$$

and

$$\nabla \cdot \mathbf{u}_t = 0 \, , \quad \nabla \cdot \mathbf{u}_\ell = \nabla \cdot \mathbf{u} \, . \qquad (7.8)$$

The equations of motion for the transverse and longitudinal components follow from the Euler–Lagrange equations as

$$\frac{\partial^2}{\partial t^2} \mathbf{u}_t = -\omega_{TO}^2 \mathbf{u}_t \tag{7.9}$$

and

$$\frac{\partial^2}{\partial t^2} \mathbf{u}_\ell = -\omega_{LO}^2 \mathbf{u}_\ell + \frac{e}{m} \mathbf{E}_{ext} \ . \tag{7.10}$$

Here, we introduced

$$\omega_{TO}^2 = \omega_0^2 \quad \text{and} \quad \omega_{LO}^2 = \omega_0^2 + \frac{4\pi e^2}{\epsilon_\infty \, v \, m} \ . \tag{7.11}$$

The equilibrium solution for the longitudinal displacement in the presence of an external field is

$$\mathbf{u}_\ell^0 = \frac{e}{m \, \omega_{LO}^2} \mathbf{E}_{ext} \ , \tag{7.12}$$

so that Eq. (7.2) yields

$$\rho_{pol} = -\frac{e^2}{vm\omega_{LO}^2} \nabla \cdot \mathbf{E}_{ext} = -\frac{4\pi e^2}{vm\omega_{LO}^2 \epsilon_\infty} \rho_{ext} \ , \tag{7.13}$$

where Maxwell's equation

$$\epsilon_\infty \nabla \cdot \mathbf{E}_{ext} = 4\pi \rho_{ext} \tag{7.14}$$

has been used. Inserting Eq. (7.13) into Eq. (7.3) and comparing the result with Eq. (7.4) allows us to identify

$$\frac{\epsilon_\infty}{\epsilon_0} = 1 - \frac{4\pi e^2}{m\omega_{LO}^2 v\epsilon_\infty} = \frac{1}{\omega_{LO}^2} \left(\omega_{LO}^2 - \frac{4\pi e^2}{m v\epsilon_\infty} \right) \ . \tag{7.15}$$

Using Eq. (7.11) shows that this is just the Lyddane–Sachs–Teller relation for the longitudinal and transverse optical frequencies

$$\frac{\omega_{LO}}{\omega_{TO}} = \sqrt{\frac{\epsilon_0}{\epsilon_\infty}} \ . \tag{7.16}$$

Before quantizing the displacement field, it is important to bring the Lagrangeian to a pure quadratic form, even in the presence of the external field. For this purpose, we subtract its equilibrium value from the displacement, which amounts to writing the lattice polarization in terms of its deviation δP from the equilibrium value. We obtain

$$\mathcal{L} = \frac{2\pi}{\epsilon_\infty} \int d^3r \left\{ \omega_{pl}^{-2} \left(\frac{\partial \delta P(\mathbf{r})}{\partial t} \right)^2 - \int d^3r' \ \delta P_\mu(\mathbf{r}) \left[\omega_0^2/\omega_{pl}^2 \ \delta_{\mu\nu} \ \delta(\mathbf{r} - \mathbf{r}') \right. \right.$$

$$\left. \left. - \frac{1}{4\pi} \partial_\mu \ \partial_\nu \ \frac{1}{|\mathbf{r} - \mathbf{r}'|} \right] \delta P_\nu(\mathbf{r}') \right\} + \frac{1}{2\epsilon^*} \int d^3r \ d^3r' \ \frac{\rho_{ext}(\mathbf{r}) \ \rho_{ext}(\mathbf{r}')}{|\mathbf{r} - \mathbf{r}'|} \ , \tag{7.17}$$

where the plasma frequency is defined through

$$\omega_{pl}^2 = \frac{4\pi e^2}{\epsilon_\infty \ v \ m} \tag{7.18}$$

and

$$\frac{1}{\epsilon^*} = \frac{1}{\epsilon_\infty} - \frac{1}{\epsilon_0} \ . \tag{7.19}$$

Now we can quantize the polarization field δP in the standard way. However, we see that we obtain a lowering of the potential energy of the state without phonons ($\delta P = 0$) in comparison to the state without external charge density $\rho_{ext}(\mathbf{r})$ by the amount

$$\frac{1}{2\epsilon^*} \int d^3r \ d^3r' \ \frac{\rho_{ext}(\mathbf{r}) \ \rho_{ext}(\mathbf{r}')}{|\mathbf{r} - \mathbf{r}'|} \equiv S \ \hbar\omega_{LO} \ . \tag{7.20}$$

Here, S is the so-called Huang-Rhys factor [Huang and Rhys (1950)].

The problem becomes more complicated if the phonons interact with a quantized electron system according to the same Coulomb scheme (Fröhlich coupling). We obtain an exactly solvable model for the electron-phonon interaction if we retain only the diagonal part of the charge density operator (commuting with the electronic Hamiltonian) and neglect the influence of the phonon interaction on the electronic states. According to our previous discussion, each electronic state $|i\rangle$ is associated with a charge density $\langle i|\rho(\mathbf{r})|i\rangle$, which adds a c-number to the phonon operators (quantization of $\delta\mathbf{P}$ instead of $\mathbf{P}$) and shifts the corresponding phonon ground-state energy according to Eq. (7.20).

If we model the LO phonons as classic oscillators, we essentially describe the energies of the coupled system through two electronic states

$$E_1(Q) = E_1 + \frac{1}{2}\hbar\omega_{LO}\,Q^2$$

$$E_2(Q) = E_2 + \frac{1}{2}\hbar\omega_{LO}\,(Q-Q_0)^2 - \frac{1}{2}\hbar\omega_{LO}\,Q_0^2\,,$$

(7.21)

where $E_{1,2}$ are the energies of the electronic system. Only the excited state has a finite local charge density.

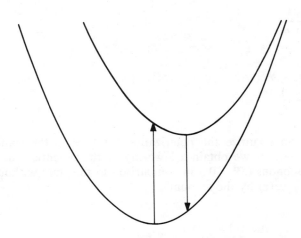

Fig. 7.1: Schematic drawing of absorption and emission in the Franck-Condon diagram.

From a Franck-Condon diagram, like the one in Fig. 7.1, it can be seen that the low temperature photon absorption occurs with a higher frequency than the photon emission. This Stokes shift is

$$\Delta = Q_0^2 \; \hbar\omega_{LO} = 2S \; \hbar\omega_{LO} \; . \tag{7.22}$$

Combining this with our previous results, we can write

$$\Delta = \frac{1}{2\epsilon^*} \int d^3r \; d^3r' \; \frac{\langle 2|\rho(\mathbf{r})|2\rangle \; \langle 2|\rho(\mathbf{r}')|2\rangle}{|\mathbf{r} - \mathbf{r}'|} \; . \tag{7.23}$$

In a quantized theory of LO phonons the result is the same, but Δ refers to the shift which is averaged over the phonon replicas. The averaging over the LO phonon replicas yields also an overall width of the spectral line of $\hbar\omega_{LO} \; 2\sqrt{S}$.

Summarizing the classical arguments, we have seen that a condition for this scheme is the existence of a strong local charge density. This implies the breaking of translational symmetry. In bulk crystals this may occur through defects (impurities) [Baroff *et al.* (1979)], or spontaneous symmetry breaking [Toyozawa (1983), Kobayashi *et al.* (1983)] like in ionic crystals.

When extending the discussion to quantum dots, one has to take into account that due to the surface polarization at the interface between the two media the effective Coulomb interaction is modified (see Sec. 2-5). This implies, that the simple Coulomb potential in the Lagrangian has to be replaced by the general potential $W_{eff}(\mathbf{r},\mathbf{r}')$ which was defined in Sec. 2-5. This general potential is a function of the background (electronic) dielectric constants of the two media and it takes into consideration in which medium either one of the two interacting particles is located. We include here the case of finite potential wells and therefore the finite probability to find particles outside the dot.

With these modifications one gets for the Stokes shift

$$\Delta = \sum_{\alpha\alpha'} \frac{\epsilon_{\infty\alpha}}{\epsilon_\alpha^*} \int_\alpha d^3r \int_{\alpha'} d^3r' \; \rho(\mathbf{r}) \; W_{eff}(\mathbf{r},\mathbf{r}') \; \rho(\mathbf{r}') \; , \tag{7.24}$$

where the summation and the corresponding integrations run over the two different media (quantum dot and its surrounding). It is important to point out that the lattice polarization charge is finite only where the charge density is non-vanishing. For neutral systems the self-energy parts of W disappear from Eq. (7.24). For the simpler case of spherically symmetric charge distributions which vanish outside the sphere of radius R, all dielectric effects of the surrounding medium disappear and we get

$$\Delta = \frac{e^2}{R\epsilon_1^*} \int_0^1 dx \int_0^x dx' \; \frac{P(x)\, P(x')}{x} \; . \tag{7.25}$$

Here $P(x) = P_h(x) - P_e(x)$ is the difference of the radial distributions of holes and electrons in the sphere and

$$x = \frac{r}{R} \; ; \quad \int_0^1 dx \; P_{e,h}(x) = 1 \; .$$

However, the dependence of the Stokes shift on the dielectric properties of the surrounding medium cannot be eliminated if the distribution inside the sphere is not spherically symmetric, or if the charge density penetrates into the medium outside the quantum dot.

The second new aspect of the quantum dot problem is that the longitudinal and transverse modes, which were introduced for an infinite medium, are no eigenmodes for phonons in a finite crystallite. The new eigenmodes and their eigenfrequencies have to be found by taking into account the boundary conditions and the symmetry of the problem. It is important to note, that for electrons confined inside the dot, only the lattice inside the dot obtains a macroscopic equilibrium deformation, even though the Coulomb interaction is long-range. Therefore it is meaningful to consider the quantization of the displacements of a continuous but finite medium. In principle, however phonons can escape from the dot, depending on the lattice matching inside and outside the dot. In the case of quantum dots embedded in glass, however, due to surface irregularities and the amorphous glassy structure, it is reasonable to think that there is no coherent phonon propagation possible between the media.

We have to take into account that our continuous medium model has a frequency-dependent dielectric constant

$$\epsilon(\omega) = \epsilon_\infty \; \frac{\omega^2 - \omega_{LO}^2}{\omega^2 - \omega_{TO}^2} \; . \tag{7.26}$$

Then the zero divergence equation for the **D**-field which defines the external sourceless eigenmodes,

$$\nabla \cdot \mathbf{D} = \epsilon(\omega) \, \nabla \cdot \mathbf{E} = 0 \tag{7.27}$$

inside and

$$\nabla \cdot \mathbf{D} = \epsilon_2 \, \nabla \cdot \mathbf{E} = 0$$

outside the sphere, has two kinds of solutions with finite field. The first solution describes oscillations with the eigenfrequency ω_{LO} and electromagnetic fields with finite divergence, whereas the second solution describes a divergence free field. This field is possible as a consequence of the boundary conditions

$$\epsilon(\omega) \, E_n \Big|_{R-0} = \epsilon_2 \, E_n \Big|_{R+0}$$

$$\tag{7.28}$$

$$E_t \Big|_{R-0} = E_t \Big|_{R+0}.$$

These conditions allow a nontrivial solution with a field for which

$$\nabla \cdot \mathbf{E} = 0$$

for any frequency that satisfies [Klein *et al*. (1990)]

$$\epsilon(\omega) = -\frac{\ell+1}{\ell} \, \epsilon_2 \quad \ell = 1,2,3 \ldots . \tag{7.29}$$

Here, we always assume that the dielectric constant ϵ_2 of the surrounding medium is frequency independent. Then Eq. (7.29) gives the eigenfrequencies of the so-called SO (surface or evanescent) modes. Such solutions do not exist in a homogeneous medium.

Since all longitudinal modes are proportional to the field, they can be expanded in the complete basis corresponding to the two kind of solutions for the potential. We have for each longitudinal mode

$$\mathbf{u(r)} \propto \nabla V , \quad (\text{ for } r < R) \tag{7.30}$$

and there are two complete sets of basis functions for the two kind of potentials. Inside the dot we have spherical Bessel functions for the LO modes

$$j_\ell(\kappa_{\ell n}\, r/R)\, Y_{\ell m}\, (\Omega) \quad (\ell = 0, 1, 2,...\ ;\quad m = -\ell,...,\ell\ ;\ n = 1, 2, 3,...)$$

with vanishing boundary conditions as well as the special homogeneous SO solutions

$$r^\ell\, Y_{\ell m}\, (\Omega) \quad (\ell = 0, 1, 2,...\ ;\quad m = -\ell,...,\ell\ )\ .$$

As in bulk materials, the number of allowed modes has to be limited so that they do not exceed the true degrees of freedom of a finite discrete lattice (Debye trick).

There is recent experimental evidence in very small ($R < a_B$) CdSe quantum dots [Alivisatos *et al.* (1988), Klein *et al.* (1990), Wang and Herron (1990)] and in $CdS_{1-x}Se_x$ quantum dots [Uhrig *et al.* (1990)] for a strong Stokes shift of the excitonic luminescence with respect to the excitonic absorption. This shift could be explained with Huang-Rhys factors S of order unity. Recent Raman-scattering experiments [Klein *et al.* (1990)] also give hints in the same direction and show evidence for the SO modes which are typical for quantum dots.

The radial charge density separation, Fig. 4.22, obtained from the spin-orbit coupling theory can be roughly described by

$$P(r) = 0.63 \sin(2\pi r) \quad (\ 0 < r < 1\ )$$

and the integrations in Eq. (7.25) can be performed easily to get for the Huang-Rhys factor

$$S_{th} \simeq 0.03\ \frac{e^2}{\epsilon^*\, R\, \hbar\omega_{LO}}.$$

Introducing the parameters of CdS or CdSe, this formula yields a Huang-Rhys factor, which is about one order of magnitude too small. Coulomb effects enhance this value, but not enough. There are still several possibilities to explain this discrepancy. First of all, the theory was built up using infinite potential barriers, which is certainly not completely correct. In a finite potential well the wavefunctions do not vanish in the neighborhood of the sphere and surface polarization effects must be taken into account. As we have seen in Chap. 4 this might lead to a substantial charge separation. Secondly, even in the frame of the simple spin-orbit coupling theory there is some freedom for speculations. The ground state of the exciton is multiple degenerate, and the wavefunctions have complicated angular dependencies. The only objective quantity is, however, the average of the charge density over all the degenerate ground states and this is radially symmetrical. A spherical symmetry-breaking may occur either due to some perturbation such as a shape irregularity, or even through energy lowering spontaneous symmetry breaking. This spontaneous symmetry breaking may again be due to the interaction with LO-phonons leading to the build-up of non radially symmetric charge distributions with stronger polarization effects. Furthermore, our analysis of the Huang-Rhys phenomenon was performed under the assumption that the lattice deformation does not affect the electronic wavefunctions. However, in a real lattice it may happen that lattice deformations, especially near the boundary, strongly affect the electronic distribution.

In this context it is worthwhile to describe here the complete self-consistent version of the Huang-Rhys phenomenon. We start from the electron-phonon Fröhlich Hamiltonian

$$\mathscr{H} = \mathscr{H}_{phon} + \mathscr{H}_{el} - \int d^3r \, d^3r' \, [\nabla \cdot \mathbf{P(r)}] \, W_{eff}(\mathbf{r,r'}) \, \rho(\mathbf{r'}) \tag{7.31}$$

and develop the phonon polarization operator and the electronic charge density around their average values in the ground state. Neglecting terms of second order in the deviations from the average values we obtain

$$\mathscr{H} \simeq \mathscr{H}_{phon} + \mathscr{H}_{el} - \int d^3r \, d^3r' \Big\{ \, [\nabla \cdot \mathbf{P(r)}] \, W_{eff}(\mathbf{r,r'}) \, \langle \rho(\mathbf{r'}) \rangle$$

$$+ \, [\nabla \cdot \langle \mathbf{P(r)} \rangle] \, W_{eff}(\mathbf{r,r'}) \, [\rho(\mathbf{r'}) - \langle \rho(\mathbf{r'}) \rangle] \Big\}. \tag{7.32}$$

We now can identify the self-consistently coupled problems of quantized phonons in the presence of the external charge and of electrons in the presence of an external polarization field. Since the divergence of the equilibrium average of the phonon polarization is given by

$$\nabla \cdot \langle \mathbf{P(r)} \rangle = \frac{\epsilon_\infty}{\epsilon^*} \langle \rho(\mathbf{r}) \rangle \tag{7.33}$$

we get for the electron system the nonlinear self-consistent Hamiltonian

$$\mathcal{H}_{eff} = \mathcal{H}_{el} - \frac{\epsilon_{\infty}}{\epsilon^*} \int d^3r \, d^3r' \, W_{eff}(\mathbf{r},\mathbf{r}') \, \langle\rho(\mathbf{r}')\rangle \, [\rho(\mathbf{r}') - \langle\rho(\mathbf{r}')\rangle] \quad . \quad (7.34)$$

This version of the theory, which in principle allows also spontaneous symmetry breaking, has not yet been studied quantitatively.

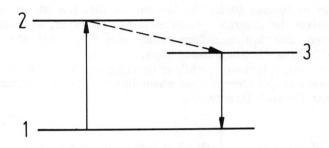

Fig. 7.2   Three-level scheme of absorption and emission.

Recently, an alternative explanation for the Stokes shift in small CdSe quantum dots was proposed [Wang and Herron (1990)] assuming, that in the excited electron-hole system one of the particles collapses into a "surface" state through interaction with the lattice. These assumption necessarily ends up again in the Huang-Rhys scheme. Indeed, in the case of a three level description (Fig. 7.2) the respective absorption and luminescence is schematically given as

$$\alpha_{12} \simeq N_2 \, |P_{12}|^2 \, \delta_{\Gamma_2+\gamma_2}(\hbar\omega - E_{21})$$

$$\alpha_{13} \simeq N_3 \, |P_{13}|^2 \, \delta_{\Gamma_3}(\hbar\omega - E_{31}) \qquad\qquad (7.35)$$

and

$$\beta_{21} \simeq n_2 \, |P_{12}|^2 \, \delta_{\Gamma_2+\gamma_2}(\hbar\omega - E_{21})$$

$$\beta_{31} \simeq n_3 \, |P_{13}|^2 \, \delta_{\Gamma_3}(\hbar\omega - E_{31}), \qquad (7.36)$$

respectively. Here $N_{2,3}$ are the degeneracies and $n_{2,3}$ the populations of the states. The matrix elements of the dipole operator determine also the radiative width of the levels

$$\tilde{\Gamma}_2 \propto |P_{12}|^2 \quad \text{and} \quad \tilde{\Gamma}_3 \propto |P_{13}|^2 \,. \qquad (7.37)$$

Generally $\tilde{\Gamma}$ (lifetime width: $\hbar/T_1$) is smaller than Γ (polarization width: $\hbar/T_2$) and γ_2 is the nonradiative decay width corresponding to the transition $2 \rightarrow 3$. The notation δ_Γ represents a broadened delta function of width Γ. At resonance we obtain

$$\alpha_{12} \propto N_2 \, \frac{\tilde{\Gamma}_2}{\Gamma_2 + \gamma_2} \quad \text{and} \quad \alpha_{13} \propto N_3 \, \frac{\tilde{\Gamma}_3}{\Gamma_3} \,. \qquad (7.38)$$

For a macroscopic crystal the degeneracies of the bulk levels are much bigger than those of the surface levels, and it may happen that one observes absorption only into the bulk level 2, whereas luminescence occurs only from the "surface" level 3, if level 2 is rapidly depopulated by nonradiative decay $2 \rightarrow 3$.

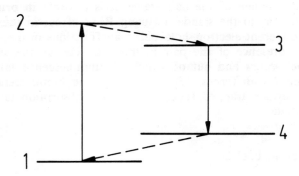

Fig. 7.3 Four-level scheme of absorption and emission.

In quantum dots the degeneracies are small and a tiny α_{13}/α_{12} ratio has no obvious explanation, at least not only in terms of the dipole matrix elements. The fact, that in experiments the $1 \rightarrow 2$ transition is observed only in absorption, while the $3 \rightarrow 1$ transition is visible only in emission, can be explained if the radiative recombination of the level 3 occurs into a fourth state, which later decays nonradiatively into the ground state 1. This four level transition scheme is shown in Fig. 7.3. Close inspection reveals, however, that also this scheme is nothing but the Huang-Rhys transition model of Fig. 7.1. It seems then plausible that the electron and hole charge densities must be significantly separated in the state 3.

A possible explanation of a strong charge separation and therefore of the big values of the Huang-Rhys factor might lie in the dielectric surface polarization effects discussed earlier. Here we think in terms of a scenario with a possible dielectric self-trapping in small quantum dots, mainly of the hole close to the surface of the dot, whereas the electron is mainly delocalized inside the dot. In this way the confinement energy roughly corresponds to the elementary kinetic confinement energy, whereas the charge distributions of the two particles are significantly separated and the probability for the hole to tunnel out of the dot into the surrounding material is increased.

An interesting possibility emerges if one takes into account that the background dielectric constant, which should be used to describe the background contributions on a short time scale, is only determined by contributions from the valence electrons (ϵ_∞), whereas on the long time scale one should use the dielectric constant ϵ_0 which also includes the slow ionic response. Since, according to the results of Sec. 4-4, a variation of the ratio ϵ of the dielectric constants inside and outside the quantum dot may result in a drastic change of the pair states, it is possible in principle to have an effect similar to the standard Huang-Rhys effect in the sense of having strongly different electron-hole-pair wavefunctions in the early and late states of the presence of the pair in the dot. Hence, the states into which absorption occurs and out of which the luminescence takes place could be substantially different. Since $\epsilon_0 > \epsilon_\infty$ the luminescence could occur from the surface trapped state, whereas the absorption takes place into the volume state.

7-2. Static Electric Fields

One of the first theoretical calculations of electric field effects [Chiba and Ohnishi (1988)] dealt with spherical GaAs dots embedded in $Al_x Ga_{1-x} As$. For this choice of materials there is a small difference of the dielectric constants and a relatively small potential barrier. Consequently, surface polarization effects can be neglected, but the finiteness of the barriers has to be taken into account. Hence, we are dealing with the one electron-hole-pair problem described by the Hamiltonian

$$\mathcal{H} = \mathcal{H}_e + \mathcal{H}_h - V_{eh} \ . \tag{7.39}$$

Here, the Hamiltonians of the electrons and holes ($\mathcal{H}_\alpha$; $\alpha = e, h$) contain the kinetic energy, the finite potential barrier, and the potential energy in the presence of the homogeneous electrical field $\mathbf{F}$,

$$\mathcal{H}_\alpha = -\frac{\hbar^2}{2m_\alpha} \nabla_\alpha^2 + U_\alpha \ \theta(r_\alpha - R) - e_\alpha \ \mathbf{F} \cdot \mathbf{r}_\alpha \ . \tag{7.40}$$

The electron–hole Coulomb interaction is written as

$$V_{eh} = \frac{e^2}{\epsilon|\mathbf{r}_e - \mathbf{r}_h|} \ . \tag{7.41}$$

Since the potential energy assumes arbitrary large negative values far enough from the dot, this problem has no stationary (discrete) states. The situation is the same as in the atomic Stark effect, where any small electric field may ionize the atom, however it will take a very long time. The correct formulation of the problem is therefore not to look for the ground-state energy, but to study the density of state of the continuous spectrum problem in the vicinity of the former bound state. For small fields one expects a spectral concentration whose shift and width increases with the applied field [see e.g. Haug and Koch (1993), Chaps. 17 and 18].

In the atomic case it is known, that the small field behavior can be predicted correctly also on the basis of lowest order stationary perturbation theory. This is the key to the success of standard textbook treatments of the atomic Stark effect in the $n = 2$ excited state. However, already the next higher order perturbation contribution is known to be wrong.

We have seen in previous chapters, that even without the complications due to the electric field, the electron–hole-pair problem in quantum dot is sufficiently complicated. Therefore, we need some drastic simplifications if we want to obtain analytic estimates for the behavior in the presence of the dc field. However, in addition to these analytic estimates we will also describe numerical results with infinite barriers, which do not need additional simplifications.

First let us discuss an approximate analysis for finite barriers. If one neglects the motion of the mass center and assumes that it always coincides with the center of the sphere [Chiba and Ohnishi (1988)], then one obtains an effective one-particle problem with the Hamiltonian

$$\mathcal{H} = -\frac{\hbar^2}{2m} \nabla^2 - \frac{e^2}{\epsilon r} - e \ \mathbf{F} \cdot \mathbf{r} + U_e \ \Theta[r - R(1 - m_e/m_h)]$$

$$+ U_h \; \Theta[r - R(1 - m_h/m_e)] \; . \tag{7.42}$$

The lack of spherical symmetry in the presence of the field suggests the use of parabolic coordinates with the z axis along the field,

$$x = \sqrt{\xi\eta} \, \cos\phi \; ; \; y = \sqrt{\xi\eta} \, \sin\phi \; ; \; z = \frac{\xi - \eta}{2}$$

$$\xi, \eta \geq 0 \; , \; 0 \leq \phi \leq 2\pi \; . \tag{7.43}$$

With the ansatz

$$\psi(\mathbf{r}) = \frac{1}{\sqrt{\xi\eta}} \, u(\xi) \, v(\eta) \, e^{im\phi} \tag{7.44}$$

separation of variables occurs if no boundary conditions exist. For the case of quantum dot the boundary problem can be handled [Chiba and Ohnishi (1988)] with another drastic approximation, which actually changes the shape of the dot from spherical to a somewhat elongated shape and modifies the barriers. Mathematically, this approximation consists in replacing the barrier potential by an effective one which has the structure

$$\frac{\xi U(\xi) + \eta U(\eta)}{\xi + \eta} \; ,$$

where the U function is the sum of two θ-functions.

After all these manipulations one may bring the problem very close to Weyl's theory of the spectral function [Weyl (1910)] and perform numerical calculations of the complex energy characterizing the unstable state. Fig. 7.4 shows the calculated energy spectrum of the exciton ground state at different field strengths. The typical shift accompanied with resonance broadening can be seen in these curves.

A simpler approach is possible for the case of infinite potential barriers. Then one has still a discrete spectrum and may apply for example variational methods to find out the ground-state energy of the electron-hole system. The only supplementary work is to take into account the dielectric effects due to the big difference of the dielectric constants for a typical quantum dot system with high barriers. Hence, one takes $U_{e,h} \rightarrow \infty$ and replaces V_{eh} in the Hamiltonian (7.39) by the more complicated Coulomb energy given in Sec. 2-3. Early investigations [Hache et $al.$ (1989), Ekimov et $al.$ (1989)] did not include the Coulomb interactions. However, Nomura and Kobayashi [(1990a) and (1990b)] used the variational wavefunctions

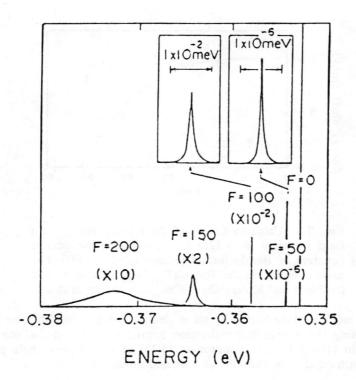

Fig. 7.4 Calculated energy spectrum of the exciton in a GaAs quantum dot for the field intensities $F =$ 50, 100, 150 and 200 kV/cm [Chiba and Ohnishi (1989)].

$$\psi \propto j_0(\pi r_e)\, j_0(\pi r_h)\, e^{-\alpha|r_e - r_h|}\, e^{\beta_e z_e + \beta_h z_h} \,, \tag{7.45}$$

where the coordinates are normalized to the sphere radius, to calculate the Stark shift of the confined ground-state exciton with and without Coulomb effects for different material parameters. In comparison to earlier calculations without Coulomb effects, the fit to the experiments on $CdS_{.12}Se_{.88}$ quantum dots is strongly improved by the calculations that include the Coulomb interaction (see Fig. 7.5), but the discrepancy for big dots still persists. This deviation may be due to the fact that even for quantum dots in glass finite barriers effects may be important.

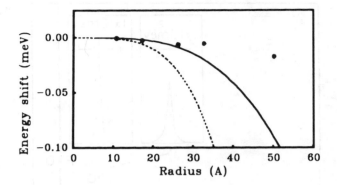

Fig. 7.5 Calculated shift of the exciton energy at a field intensity $F = 12.5 \, kV/cm$ (inside the dot) as function of the dot radius [Hache *et al.* (1989)]. The experimental results for $CdS_{.12}Se_{.88}$ quantum dots [Nomura and Kobayashi (1990a)] are shown as dots.

The modifications caused by static electric fields have also been investigated using the matrix diagonalization approach of Sec. 4-2 [Esch *et al.* (1990), Hu (1991)]. The Hamiltonian (4.20) of the electron–hole pair in the dot with rigid walls contains now an additional piece, i.e.,

$$\mathscr{H}_x \rightarrow \mathscr{H}_x - e \, \mathbf{F} \cdot (\mathbf{r}_e - \mathbf{r}_h) \, . \tag{7.46}$$

This new Hamiltonian is invariant only against rotations around the z-axis defining the field direction. Even though the total angular momentum is no longer a conserved quantity, its z-component is still conserved. The field term, which is a vector operator, raises or lowers the angular momentum by one. This leads to the mixing of exciton states with different angular momenta. One may still use the Wigner-Eckart theorem for irreducible tensors belonging to the $L = 1$ representation of the SU_2 group (vector operator)

$$\langle \ell,m | \, T_M^{(1)} \, | \ell',m' \rangle = \langle \ell,M;\ell',m' | \ell,m \rangle \, (\ell | T^{(1)} | \ell') \tag{7.47}$$

in order to reduce the number of independent matrix elements. In other words

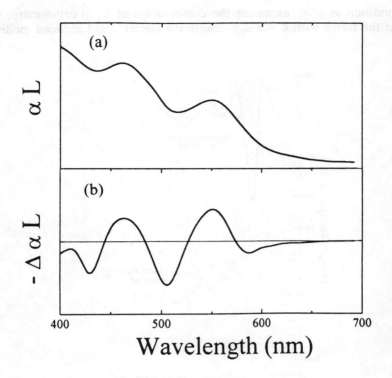

Fig. 7.6: Calculated linear absorption (a) and electric field induced change in absorption (b) for CdS quantum dots.

$$\langle n,\ell,m |\ r\cos\theta\ | n',\ell',m' \rangle$$

$$= \int dr\ r^3\ \phi_{n\ell}(r)\ \phi_{n'\ell'}(r)\ (A_{\ell'm}\,\delta_{\ell'+1,\ell} + A_{\ell'-1,m}\,\delta_{\ell'-1,\ell})\ \delta_{mm'}, \qquad (7.48)$$

where

$$A_{\ell,m} = \sqrt{\frac{(\ell+1)^2 - m^2}{(2\ell+1)\,(2\ell+3)}}\ . \qquad (7.49)$$

The condition $m = m'$ expresses the conservation of L_z. Furthermore, we see that the terms with $\ell' = \ell \pm 1$ imply the electric field induced modifi-

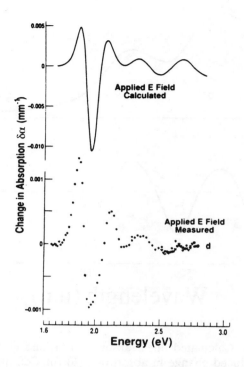

Fig. 7.7: Calculated (full curve) and measured (dots) change of absorption in CdTe Quantum Dots due to the introduction of a static field. From Esch *et al.* (1990).

cations of the dipole transition rules.

Using Eqs. (7.46) - (7.49) the matrix diagonalization technique has been applied to evaluate the optical susceptibility as discussed in Chap. 5. Using the example of CdS quantum dots, we compare the computed linear absorption spectrum in the presence of an external electric field with the linear absorption spectrum without field. Fig. 7.6 shows the linear absorption and the field induced absorption changes.

For the case of CdTe dots with $R/a_B = 0.5$ in glass, the variation of the electron-hole-pair absorption in a uniform external field of 58 kV cm^{-1} was measured [Esch *et al.* (1990)]. Fig. 7.7b shows that the measured absorption changes are in good qualitative agreement with theoretical calculations made assuming infinite barriers.

7-3. Magnetic Fields

Special interest exists for disk shaped dots (quantum disks) with soft barriers which are realized through ion implantation. The barriers in such structures are believed to be well approximated through parabolic potentials. The high degree of symmetry in the presence of a homogeneous magnetic field applied perpendicular to the plane of the dot allows exact solutions of the one-particle motion and a very simple explanation of measured infrared electronic absorption spectra. At the same time, these quantum disks stimulated extensive calculations of bounded two-dimensional few electron systems. Many of these topics are closely related [Sivan *et al.* (1989)] to the famous quantum Hall effect, which is a property of two-dimensional electron systems.

Let us consider the Hamiltonian of N two-dimensional Coulomb interacting electrons in the presence of a magnet field **B** oriented along the z axis, perpendicular to the plane of electronic motion. The quantum disk is defined by the existence of an oscillator potential in the plane.

$$\mathcal{H} = \sum_{i=1}^{N} \mathcal{H}_i + \frac{1}{2} \sum_{\substack{i,j \\ (i \neq j)}}^{N} \frac{e^2}{\epsilon |\mathbf{r}_i - \mathbf{r}_j|} \qquad (7.50)$$

where the Hamiltonian of each electron $(i = 1,\dots,N)$ is

$$\mathcal{H}_i = \frac{\hbar^2}{2m} [i\hbar \, \nabla_i - e \, \mathbf{A}(\mathbf{r}_i)]^2 + \frac{1}{2} m \, \omega_0^2 \, \mathbf{r}_i^2 \,. \qquad (7.51)$$

The frequency ω_0 of the oscillator potential is inversely proportional to the radius of the two-dimensional dot $(\omega \propto 1/R)$.

The vector potential may be chosen as

$$\mathbf{A}(\mathbf{r}) = \frac{1}{2} \, \mathbf{B} \times \mathbf{r} \,. \qquad (7.52)$$

The neutralizing positive background is supposed to be situated outside the dot and gives only a constant energy shift which is ignored since no compensation of divergences is needed in a finite system. The spin degree of freedom is also ignored, assuming that the field **B** is strong enough to align all spins.

First of all we want to show, that the one electron problem is exactly solvable [Fock (1928), Darwin (1930), Dingle (1952)]. Since the Hamiltonian (7.52) is invariant with respect to rotations around the z axis through the center of the disk, one may write the one-electron wavefunctions as

$$\psi(\mathbf{r}) = u_M(r) \, e^{iM\phi} . \tag{7.53}$$

Then u_M obeys the radial Schrödinger equation

$$\left[-\frac{\hbar^2}{2m} \left(\frac{\partial^2}{\partial r^2} + \frac{1}{r} \frac{\partial}{\partial r} \right) + \frac{M^2 \hbar^2}{2mr^2} + \frac{m}{8}(\omega_c^2 + 4_0^2)r^2 - \frac{1}{2} M \hbar\omega_c \right] u_M = E \, u_M , \tag{7.54}$$

where $\omega_c = eB/mc$ is the cyclotron frequency. The solutions of Eq. (7.54) are

$$u_M(r) = r^{|M|} L_n^{|M|}(r^2/2a^2) \, e^{-r^2/4a^2} , \tag{7.55}$$

where $L_n^{|M|}$ is a Laguerre polynomial and

$$a^2 = \frac{\hbar}{m} (\omega_c^2 + 4\omega_0^2)^{-1/2} . \tag{7.56}$$

The energy eigenvalues are

$$E_{nM} = (2n + 1 + |M|) \, \hbar \, (\omega_c^2/4 + \omega_0^2)^{1/2} - \frac{1}{2} M \hbar\omega_c . \tag{7.57}$$

In the limit without confinement ($\omega_0 \to 0$) one recovers the Landau solution where the energy depends only on the quantum number

$$\mathscr{N} = n + \frac{|M| - M}{2} . \tag{7.58}$$

Without the confining potential the energies of the positive M states would be independent of M, but in the presence of quantum confinement they increase with M. This is the key difference between the behavior of free and confined electrons and it is responsible for much of the new physics. The degeneracy of the bulk Landau levels is removed by the confining potential which varies smoothly in space. By contrast, sharp boundary conditions leave most of the bulk levels degenerate, only lifting the degeneracy of few states at the edges. From this point of view all eigenstates in our case are edge states.

The states of a system of $N = 3$ and $N = 4$ electrons were calculated [Maksym and Chakraborty (1990)] by numerically diagonalizing the many electron Hamiltonian (7.50). The unusual feature emerges that the ground state angular momentum J increases with B, in contrast to the case of non-interacting electrons where the ground state has the lowest available total J, provided B is so high that only $\mathcal{N} = 0$ is relevant. Surprisingly however, the ground state in the quantum disk does not take all possible values of J.

The rich structure of the calculated spectra is somewhat in contrast with current experimental results on far-infrared spectroscopy [Milanovic and Ikonic (1989), Sikorski and Merkt (1990)]. The experimental results show just two features whose energies seem to correspond to single-electron excitations, independent of the number of electrons in the dot. This difference can be understood [Maksym and Chakraborty (1990), Peeters (1990), Bakshi et al. (1990)] if one takes into account that the interaction with electromagnetic radiation, which has a wavelength that is large in comparison to the dot size, is described by the dipole Hamiltonian

$$\mathcal{H}'(t) = \sum_{i=1}^{N} e\, \mathbf{E_0} \cdot \mathbf{r}_i\, e^{i\omega t} . \tag{7.59}$$

The interaction, hence, depends only on the sum of all the coordinates and can therefore be written in terms of the center of mass (cm) coordinate

$$\mathbf{R} = \frac{1}{N} \sum_{i=1}^{N} \mathbf{r}_i \tag{7.60}$$

and the total charge $Q = Ne$,

$$\mathcal{H}'(t) = Q\, \mathbf{E_0} \cdot \mathbf{R}\, e^{i\omega t}. \tag{7.61}$$

To understand the consequences, it is convenient to rewrite the many electron Hamiltonian (7.50) in terms of the cm and $2(N-1)$ relative coordinates. These are most conveniently chosen to be

$$\mathbf{r}'_i = \mathbf{r}_i - \mathbf{R} \quad (i = 1, ..., N-1) . \tag{7.62}$$

Due to the quadratic form of the confinement potential the many-electron

Hamiltonian admits then a separation of variables, like in the known case without confinement

$$\mathscr{H} = \frac{\hbar^2}{2Nm} [i\hbar \, \nabla_R + Q \, \mathbf{A(R)}]^2 + \frac{1}{2} N \, m \, \omega_0^2 \, R^2 + \mathscr{H}_{rel} \, . \qquad (7.63)$$

The last term is only a function of the relative coordinates and contains all the effects of the interaction. The cm motion is the only one which is coupled to the electromagnetic field. This motion is identical to that of a single particle of mass $N\,m$ and charge $N\,e$ in the parabolic confinement potential. Since the cyclotron resonance frequency depends only on the charge to mass ratio, its spectrum coincides with that given by Eq. (7.57) and is therefore not sensible to the number of electrons in the system. Far infrared absorption experiments in the parabolic confinement see only features at the single-electron energies [Sikorski and Merkt (1989)].

Other recent theoretical investigations [Kumar et al. (1990), Lent (1991)] of confined two-dimensional few electron systems in magnet field consider different confinement potentials corresponding also to realistic gate configurations [Kumar et al. (1990)] in the frame of the self-consistent potential approximation.

Chapter 8
COUPLED QUANTUM DOTS

In previous chapters we analyze the properties of a quantum-dot system by computing the properties of a single dot and averaging over a dot-size distribution. Such an approach can be well justified for semiconductor clusters embedded in glass or for colloidal solutions, as long as the distances between neighboring dots are large and random. However, for quantum dot systems which are prepared through etching of two-dimensional quantum wells a well defined correlation of the dot positions is possible. The simplest situation of such regular quantum dots corresponds to two closely separated dots between which carrier tunneling is possible [Bryant (1993)].

In addition to systems with a few coupled dots it is also possible to pattern two-dimensional layers to create entire quantum dot superlattices [Reed *et al.* (1988) and (1990), Kern *et al.* (1991), Demel *et al.* (1990a) and (1990b), Lorke *et al.* (1990), Kastner (1993), Reed (1993)]. The development of these and similar elaborate structures clearly make it worthwhile to invest some theoretical effort into the study of coherent properties of arrays of quantum dots (disks) in which the dots themselves are only schematically characterized.

For dilute quantum dots embedded in glass carrier tunneling from the dot is also possible, however, usually not to other dots but to localized levels which are distributed randomly in the gap of the glass material. These glass levels may be due to impurities or they may be intrinsic localized levels due to the disordered amorphous structure of the glass. Although the probability to find a resonating level in the vicinity of the dot. The levels in the glass may be due to impurities or they may be intrinsic localized levels due to the disordered amorphous nature of the glass. Although the probability to find such a level in resonance with a quantum dot level is small, the transition might occur under the assistance of phonons. Through such incoherent tunneling processes an particles are transferred irreversibly from the dot to its surrounding.

Incoherent nonradiative transitions might occur also to defect states in the quantum dot itself or to states typical for the surface of the dot. These processes are the background for the explanation of the glass-darkening (photo darkening), which occurs after strong illumination above the absorption threshold of the quantum dot [Roussignol *et al.* (1987)].

8-1. Double Quantum Dots

A schematic arrangement of a double dot system is shown in Fig. 8.1. Two quasi-two-dimensional GaAs boxes and a separating AlGaAs box form a coherent two-dot system. For a sufficiently thin AlGaAs barrier layer carrier tunneling occurs between the two quantum dots.

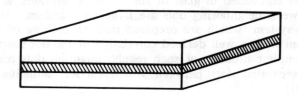

Fig. 8.1: Schematic drawing of a double quantum dot system. The top and bottom layer are the two quantum boxes which are separated by a barrier material symbolized by the shaded area.

In the theoretical discussion of such a double quantum dot system it is usually assumed [Bryant (1993)] that all the confining potential barriers may be taken to be infinite, with the exception of the barrier that couples the two dots. The finite intermediate barrier corresponds to the band off-sets of the composite materials. Fig. 8.2 shows the effective potential for an electron (or hole) in the direction perpendicular to the layer structure (z-direction). The lateral dimensions of the two boxes are identical but their thickness is not necessarily the same. In this example the upper box is thicker.

In the absence of the Coulomb interaction the properties of the double dot system may be schematically described in terms of a two-level model. One considers the lowest lying one-particle states of the independent dots $|1\rangle$, $|2\rangle$, the corresponding energies ϵ_1, ϵ_2 and a tunneling amplitude t between the two dots. The effective Hamiltonian for one of the particles is

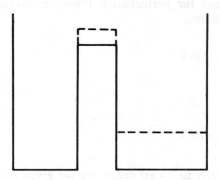

Fig. 8.2: The potential of a double quantum dot system in the transverse direction. In this example the upper box (right part of the potential) is thicker. The dashed line is the potential in the presence of a bias field.

$$\mathscr{H} = \epsilon_1 \, |1\rangle \langle 1| + \epsilon_2 \, |2\rangle \langle 2| + t \, |1\rangle \langle 2| + t^* \, |2\rangle \langle 1| , \tag{8.1}$$

where the parameters are usually different for electrons or holes.

The quantized one-particle wavefunctions are just the sum of the two quantized one-particle wavefunctions in the two dots

$$\psi_\alpha(\mathbf{x}) = \psi_{\alpha 1}(\mathbf{x}) + \psi_{\alpha 2}(\mathbf{x}) , \qquad (\alpha = e, h) . \tag{8.2}$$

Since the wavefunctions in the independent dots have no overlap (all the tunneling effects are in the tunneling parameter t), the dipolar polarization operator in terms of the wavefunctions (8.2) is given by

$$P = p_{cv} \int d\mathbf{x} \, \psi_e(\mathbf{x}) \, \psi_h(\mathbf{x}) + \text{h.c.}$$

$$= p_{cv} \int d\mathbf{x} \, [\psi_{e1}(\mathbf{x}) \, \psi_{h1}(\mathbf{x}) + \psi_{e2}(\mathbf{x}) \, \psi_{h2}(\mathbf{x})] + \text{h.c.} . \tag{8.3}$$

For the case of symmetrical dots ($\epsilon_1 = \epsilon_2$) we have degenerate levels, and the suitable basis for perturbation theory is given by the symmetric and antisymmetric states,

$$|s\rangle = \frac{1}{\sqrt{2}}(|1\rangle + |2\rangle)$$

and   (8.4)

$$|a\rangle = \frac{1}{\sqrt{2}}(|1\rangle - |2\rangle).$$

The perturbed energies to lowest order in t are given by

$$E_s = \epsilon - t$$
                                                                                       (8.5)
$$E_a = \epsilon + t.$$

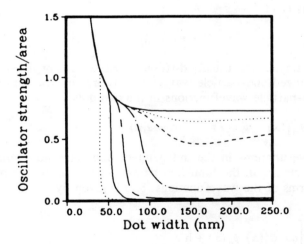

Fig. 8.3: Ground state oscillator strength (normalized to the dot area) as function of dot width at different bias: 0 meV (upper solid curve), 7.7 mev (upper dotted curve) 7.9 meV (dash), 8.1 meV (dash-dot), 8.5 meV (long-short dash), 9 meV (lower solid curve), 10 meV (lower dotted curve). From Bryant (1993).

The tunneling parameter t has to be chosen real and positive to correspond to the original model in real space. The Hamiltonian is symmetrical with respect to the permutation of the two dots and the ground state has to obey this symmetry. It is easy to see that, within this approximation, recombination might occur only if the electron and the hole are in the same symmetrical or antisymmetric state, whereas the mixed transitions are dipole forbidden. However, a symmetry breaking applied electric field allows new transitions.

The whole picture becomes much more complicated as soon as one includes the electron-hole Coulomb interaction. The role of this interaction is enhanced by the lateral quantum confinement in comparison to the analogous quantum well problem. Bryant (1993) performed a variational calculation in real space including the Coulomb interaction for the asymmetric transverse potential configuration shown in Fig. 8.2. His physical parameters correspond to the system $GaAs/Al_{0.3}Ga_{0.7}As/GaAs$. The calculations were performed also in the presence of a bias, distributed as shown in Fig. 8.2 by the dotted lines. Simultaneously with the wavefunctions and their energies the oscillator strengths were calculated. For fixed bias the numerical results (Fig. 8.3) show a strong dependence of the ground state oscillator strength (normalized to the box area L^2) on the lateral confinement L.

8-2. Quantum Dot Superlattices

As mentioned earlier, it is technologically possible to prepare two-dimensional lattices of quantum dots as in Fig. 8.4a, or one dimensional arrays of dots as shown in Fig. 8.4b. Even three dimensional arrangements can be obtained through successive layering.

The qualitative difference between layered superlattices with one-dimensional periodicity and two-dimensional superlattices of quantum dots can be seen easily in the zone-folded band structure for a free electron in a two-dimensional layer with imposed periodicity of $A_x = A_y = 30$ nm as shown in Fig. 8.5. In a one-dimensional superlattice, zone folding occurs only in one direction and only the $k_y = 0$ band is present. When a two-dimensional periodicity is imposed additional doubly degenerate zone-folded bands appear. These new bands are replicas of the $k_y = 0$ band shifted to higher energy by $E(k_x = 0, k_y = 2\pi n/A_y)$. The possibilities for band modifications are substantially enhanced for the case of imposed two-dimensional periodicity, because then a single band can have multiple crossings with its zone-folded replicas. It should be noted that in the presence of a weak periodic potential all the crossing points will develop band gaps through the lifting of the degeneracy. This superlattice band structure (for the given lattice constant $A = 30$ nm) has an energy width in the meV range.

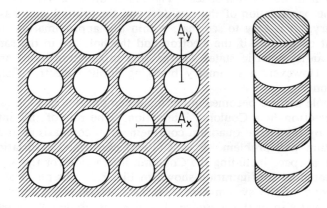

Fig. 8.4: One- and two-dimensional lattices of quantum disks.

With a little fantasy one may envisage several kinds of artificial two-dimensional superlattices built up from regions with materials having different band off-sets. Examples of the resulting potential well configurations are shown schematically in Fig. 8.6. In the box configuration most of the charge concentrates in the dot, in the anti-dot it will flow in the inter-dot channels, whereas the resonator configuration may lead to a mixture of both situations. Bryant (1989b) performed extended augmented-plane-wave calculations for such structures. The mini-bands and the corresponding resonant tunneling were also studied theoretically [Aers and Liu (1990)] and experimentally [Kouwenhoven *et al.* (1990)].

Kayanmua (1993) recently analyzed the resonant interaction of photons with quantum-dot arrays. He finds polariton like modes for regular dot arrangements and a transition to random Rayleigh scattering for irregularly arranged dots.

Another kind of interesting collective property of quantum dot arrays is the Coulomb many-body charge coupling between the dots [Que and Kirczenow (1988)]. This problem has been treated in the frame of the self-consistent random phase approximation (RPA) to show the existence of tunneling plasmon oscillations characteristic for arrays of quantum dots [Zhu *et al.* (1989), Huang and Zhou (1989), Huang and Antoniewicz (1991)]. One might expect to see the computed frequencies in Raman scattering experiments.

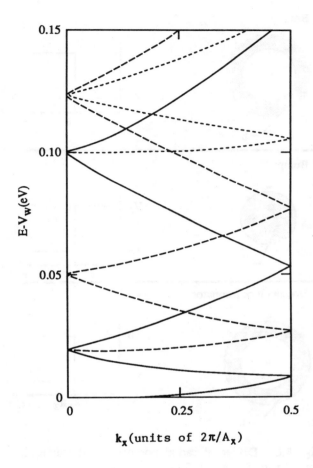

Fig. 8.5: Bandstructure of free two-dimensional electrons. The k_x dispersion for $k_y = 0$ is plotted as solid curves, for $k_y = \pm 2\pi/A_y$ as dashed curves, and for $k_y = \pm 4\pi/A_y$ as dotted curves, respectively. For the calculations $A_x = A_y = 30$ nm was chosen. From Bryant (1989b).

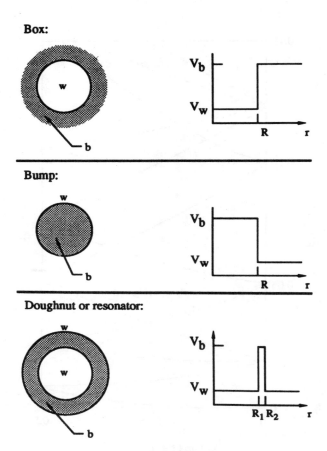

Fig. 8.6: Different radial potential well configurations. From Bryant (1989b).

8-3. Quantum Dots in a Random Matrix

For the case of semiconductor quantum dots embedded in glass one has to take into account that the glass itself has an amorphous structure and therefore its gap is actually only a so called *mobility gap*. The *mobility edge* separates the delocalized (mobile) states with energy larger than the mobility gap from the localized states inside the gap [Bányai (1964), Mott (1967)]. Mobility of the carriers in the localized states is possible only with the help of phonon absorption and emission. States which are close in real space are typically far away in energy. Through phonon assisted hopping [Mott (1968)] this obstacle is overcome, but the mobility decreases rapidly with decreasing temperature. As discovered by Anderson (1958), the localized states are mostly due to an intrinsic destructive interference induced by disorder (Anderson localization).

Real amorphous structures are characterized also by structural defects (voids) and unsatisfied valence bonds (dangling bonds). At the same time also true impurity states occur in abundance. For example, such impurity

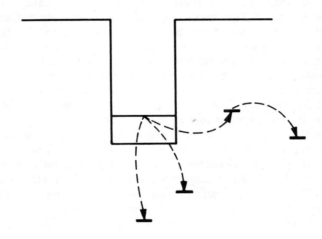

Fig. 8.7: Incoherent transitions from a state inside a quantum dot.

states may be due to uncrystallized dopant material left in the glass matrix.

All these complicated features make a quantitative theoretical description very difficult. Nevertheless, some qualitative conclusions can be drawn. The carriers created through absorption in the quantum dot may leave the dot through phonon assisted (incoherent) transitions. First they will be localized close to the quantum dots, but eventually they might diffuse away. The quantum dot system under laser illumination therefore acts as a carrier source for the glass, which is otherwise non-absorbing in this spectral region. Rapid nonradiative transitions might occur to strongly localized states on the surface of the dot (e.g. dangling bonds) or deep states inside the dot.

All these one-particle incoherent couplings (see Fig. 8.7), with possible many-particle effects in the case of double occupations, are at the origin of the irreversible optical phenomena in the semiconductor quantum dot systems in a glass matrix. These phenomena are generically denoted as photo darkening of the glass.

Trapping states and photo darkening have been observed e.g. in $CdS_x Se_1$-x doped glasses [Roussignol et al. (1987), Williams (1988), DeLong et al. (1989), Tomita and Matsuoka (1990), Ma et al. (1992), Schanne-Klein et al. (1992)]. This irreversible process is characterized after laser annealing by (i) a decrease in luminescence intensity, (ii) a decrease in the intensity of nonlinear optical signals, and a shortening of the nonlinear-optical response time. (iii) These glasses show permanent holographic gratings, but (iv) photo darkening only weakly affects the absorption spectra. (v) The darkening rate decreases substantially at low temperatures. Host glass and not heat treated samples exhibit little or no photodarkening. A glass that suffered this change recovers its initial condition after annealing for several hours at elevated temperatures.

Phenomenological descriptions of these properties based on the incoherent couplings discussed at the beginning of this section are possible [Tomita and Matsuoka (1990), Malhotra (1988)]. The crystallite surface might very well play a special role in these processes.

Additional insights are gained through experiments made on quantum dots (made out of the same material) embedded in organic materials [Bawendi et al. (1990), Alivisatos et al. (1988)]. Recently it was found [Woggon et al. (1993) that quantum dots embedded in glass show, after strong laser illumination, spectrally narrow hole-burning, similar to the results reported for organic matrices. This was tentatively attributed to changes of interface charge states or interface polarizations under high excitation.

Appendix
ASYMPTOTIC CLUSTER GROWTH LAWS

Under ideal conditions, the size distribution and the rate of growth of quantum dots are direct consequences of the physical growth mechanism. These features have been analyzed for the kinetics of precipitation from supersaturated solid solutions by Lifshitz and Slezov (1959, 1961). The Lifshitz-Slezov results have been generalized for different growth conditions by Desai et al. (1983), see also Koch et al. (1983) and Koch (1984). The asymptotic cluster growth law is found to be a direct consequence of the rate with which monomers from the surrounding medium arrive at the cluster surface. In this appendix, we summarize the main ideas and results of such an asymptotic analysis of the cluster growth dynamics.

A-1. Dynamic Equations

The kinetic equation for an individual cluster can be written as the rate equation for the number of particles N in a cluster

$$\frac{dN}{dt} = g_N - l_N . \tag{A.1}$$

Here g_N is the "gain rate", i.e. the rate at which monomers (atoms or molecules) from the surrounding medium condense into the cluster. Correspondingly, l_N is the "loss rate", i.e., the rate of monomers lost from the cluster to the surrounding medium. In order for the cluster to grow, g_N has to be larger than l_N, which can only happen if the surrounding medium is supersaturated, which means that it has an effective monomer density larger than the thermodynamic equilibrium density for coexistence between monomers and cluster.

The detailed form of the loss rate can be obtained as the thermionic emission rate of monomers through the cluster surface,

$$l_N = v A \, n_T(R) \; ,$$
(A.2)

where v is the thermal velocity, n_T is the thermal density,

$$n_T(R) = n_T \, e^{\sigma/R} \; ,$$
(A.3)

and

$$A = 4\pi \, R^2 \; ,$$
(A.4)

is the surface area of the spherical cluster of radius R. The density $n_T \propto \exp(-\phi/T)$ where ϕ is the volume energy, T is the temperature, and the factor $\exp(\sigma/R)$ in Eq. (A.3) is the surface contribution, where σ is proportional to the surface tension of the cluster.

The gain rate is the current of monomers through the cluster surface,

$$g_N = A \, v \, n_v(R) \; ,$$
(A.5)

where n_v is the monomer density in the surrounding medium just outside the cluster surface. Here we assume that the overall cluster density is low so that each cluster interacts directly only with its surrounding medium and not with other clusters. Indirect interaction between clusters is, however, possible since a small cluster may evaporate and the monomers may travel through the matrix medium to the large cluster. In this sense the matrix material with all the other clusters appears as an effectively supersaturated surrounding for each individual cluster.

Using Eqs. (A.2), (A.5), and

$$N = \frac{4}{3} \pi \, n_0 \, R^3 \; ,$$
(A.6)

with n_0 being the density inside the cluster, we can transform Eq. (A.1) into an equation for the cluster radius

$$\frac{dR}{dt} = \frac{v}{n_0} \left[n_v(R) - n_T(R) \right] .$$
(A.7)

To compute the monomer density at the cluster surface, we have to specify the process how the monomers come to the cluster. In the classical Lifshitz–Slezov analysis one assumes that the monomers diffuse through the surrounding medium. Hence, one obtains the density from the diffusion equation

$$\frac{\partial}{\partial t} n_v = D \nabla^2 n_v \; ,$$

(A.8)

where D is the diffusion coefficient. Under quasi-stationary conditions, Eq. (A.8) can be written as

$$D \; \frac{1}{r^2} \frac{\partial}{\partial r} r^2 \frac{\partial}{\partial r} n_v = 0 \; ,$$

(A.9)

where we assumed radial symmetry. The boundary condition for the diffusion current at the cluster surface is

$$D \left. \frac{\partial n_v}{\partial r} \right|_R = v \left[n_v(R) - n_T(R) \right] .$$

(A.10)

The solution of Eq. (A.9) with this boundary condition is

$$n_v(r) = n - R^2 \, v \; \frac{[n_v(R) - n_T(R)]}{Dr} \; ,$$

(A.11)

where n is the average density well outside the cluster. Solving Eq. (A.11) for $r = R$ yields

$$n_v(R) = \frac{1}{1 + Rv/D} \left[n + \frac{Rv}{D} n_T(R) \right] .$$

(A.12)

Inserting this result into Eq. (A.7) for the cluster radius gives

$$\frac{dR}{dt} = \frac{v}{n_0(1 + Rv/D)} \left[n - n_T(R) \right] .$$

(A.13)

Since we are interested in the growth of larger clusters we ignore the 1 in the denominator of Eq. (A.13) and expand

$$n_T(R) \simeq n_T \left(1 + \frac{\sigma}{R} \right)$$

(A.14)

to obtain

$$\frac{dR}{dt} = \frac{D}{R}\left(\Delta - \frac{\alpha}{R}\right), \tag{A.15}$$

where

$$\Delta = \frac{n - n_T}{n_0} \tag{A.16}$$

is the normalized supersaturation of the medium outside the cluster and

$$\alpha = \frac{n_T \sigma}{n_0}. \tag{A.17}$$

To simplify the analysis, we transform the equations to scaled variables. First of all, we introduce a "critical droplet radius" from Eq. (A.15),

$$R_c(t) = \frac{\alpha}{\Delta(t)}. \tag{A.18}$$

Droplets with a size smaller than R_c shrink whereas the larger ones grow. Note, however, that Eq. (A.18) does not necessarily yield the physically correct critical droplet size, since Eq. (A.15) has been obtained for large R, and the critical size may be small. Here, we use $R_c(t)$ only as a convenient measure for the supersaturation. If we would want to study critical droplets we would use the full Eq. (A.13).

Introducing

$$\rho(t') = R(t')/R_c(0), \quad t' = t\alpha D/R_c^3(0),$$

$$x(t') = \Delta(0)/\Delta(t') = R_c(t')/R_c(0), \tag{A.19}$$

Eq. (A.15) can be written as

$$\frac{d\rho^3}{dt'} = 3\left(\frac{\rho}{x} - 1\right). \tag{A.20}$$

To obtain an equation for the supersaturation, we use the conservation of the total number of monomers. We denote by Q_0 the total supersaturation at time zero. This supersaturation in the sum of the supersaturation in the medium outside the clusters, Δ_0, plus all the monomers q_0 which are originally already inside clusters,

$$Q_0 = \Delta_0 + q_0 = \Delta(t') + q(t') . \qquad (A.21)$$

Since the total number of monomers is conserved, Eq. (A.21) has to hold for all times.

We write the amount of material in all clusters as first moment of the droplet volume distribution function $f(\rho^3, t')$,

$$q(t) = \frac{4}{3} \pi R_c^3(0) \int_0^\infty d\rho^3 \, \rho^3 \, f(\rho^3, t') . \qquad (A.22)$$

The integral in Eq. (A.22) is proportional to the total number of monomers in all the clusters. The distribution function has been normalized such that,

$$\int_0^\infty d\rho^3 \, f(\rho^3, t') = n_{cluster}, \qquad (A.23)$$

where $n_{cluster}$ is the density of clusters per unit volume, and

$$q_0 = \frac{4}{3} \pi R_c^3(0) \int_0^\infty d\rho^3 \, \rho^3 \, f(\rho^3, t'=0) . \qquad (A.24)$$

To compute the volume distribution function, we use the continuity equation for the probability density,

$$\frac{\partial}{\partial t'} f(\rho^3, t') + \frac{\partial}{\partial \rho^3} J(\rho^3, t') = 0 . \qquad (A.25)$$

The probability current is

$$J(\rho^3, t') = f(\rho^3, t') \, v_\rho , \qquad (A.26)$$

where $v_\rho = d\rho^3/dt'$ is the cluster volume growth rate.

The coupled equations (A.20), (A.21), and (A.25) are the starting point of the asymptotic analysis which yields the cluster growth law and the cluster distribution.

A-2. Scaled Equations

For the further analysis it is helpful to make another variable transformation. We define

$$z = \rho^3/x^3 \quad \text{and} \quad \tau = \ln x^3 , \tag{A.27}$$

i.e., we measure the time through the increasing critical droplet radius R_c. Inserting (A.27) into (A.20) yields

$$\frac{dz}{d\tau} = \gamma(\tau) \ (z^{1/3} - 1) - z \equiv \nu(z,\gamma) , \tag{A.28}$$

where

$$\gamma(\tau) = \frac{3}{\dfrac{dx^3}{dt'}} . \tag{A.29}$$

We now scale the particle number conservation equation (A.18),

$$1 = \frac{\Delta(t')}{Q_0} + \kappa \int_0^\infty d\rho^3 \ \rho^3 \ f(\rho^3,t')$$

$$= \frac{\Delta_0}{Q_0} \ e^{-\tau/3} + \kappa \ e^\tau \int_0^\infty dz \ z \ \phi(z,\tau) , \tag{A.30}$$

where

$$\kappa = \frac{4}{3} \ \pi \ \frac{R_c^3(0)}{Q_0} . \tag{A.31}$$

In Eq. (A.30) we made the transformation

$$d\rho^3 f(\rho^3,t') \ \rightarrow \ dz \ \phi(z,\tau) \tag{A.32}$$

and used Eqs. (A.19) and (A.27) to express the normalized supersaturation in terms of τ.

Let us assume that

$$z(\tau) = z(y,\tau) \tag{A.33}$$

is a solution of Eq. (A.28) which fulfills the initial condition

$$z(0) = y . \tag{A.34}$$

Since we treat the motion as deterministic, the probability distribution at any time may be expressed through the probability distribution of the initial values. Therefore, the integral in Eq. (A.30) may be written in terms of the initial distribution function

$$\phi(z,\tau=0) \equiv f_0 [z(y)] \equiv f_0(y) . \tag{A.35}$$

Using

$$dz \ \phi(z,\tau) = dy \ f_0(y) , \tag{A.36}$$

Eq. (A.30) becomes

$$1 = \frac{\Delta_0}{Q_0} e^{-\tau/3} + \kappa \ e^{\tau} \int_{y_0(\tau)}^{\infty} dy \ z(y,\tau) f_0(y) , \tag{A.37}$$

where $y_0(\tau)$ is the solution of

$$z[y_0(\tau),\tau] = 0 . \tag{A.38}$$

Finally, Eq. (A.25) in scaled form is

$$\frac{\partial}{\partial \tau} \phi(z,\tau) + \frac{\partial}{\partial z} [\ \phi(z,\tau) \ v(z,\gamma) \ ] = 0 . \tag{A.39}$$

A-3. Asymptotic Analysis

We now use asymptotic analysis to determine the unknown functions $\gamma(\tau)$ and $\phi(z,\tau)$. For this purpose we inspect

$$\frac{1}{\gamma(\tau)} = \frac{1}{3} \frac{dx^3}{dt'} = x^2 \frac{dx}{dt'} . \tag{A.40}$$

Since x is inversely proportional to the supersaturation, Eq. (A.19), which decreases with increasing time because of the growing clusters, we know that

$$\frac{dx}{dt'} > 0 , \tag{A.41}$$

and, hence, $\gamma > 0$. For $\tau \to \infty$, γ can converge toward one of three possible values,

a) $\gamma \to \infty$ (A.42a)

b) $\gamma \to 0$ (A.42b)

c) $\gamma \to \gamma_0 = $ const.   (A.42c)

To see which asymptotic value of γ is realized we check the compatibility of (A.42a) - (A.42c) with the matter conservation equation (A.30) or (A.37).

a) $\gamma \to \infty$

This case would imply

$$\frac{dx^3}{dt'} \to 0 , \tag{A.43}$$

so that x^3 would grow slower than t'. To check the consistency of this case we look at the dynamic evolution of all monomers in all clusters. For this purpose we take the first moment of the continuity equation (A.39),

$$\frac{\partial}{\partial \tau} \int_0^\infty dz \, z \, \phi(z,\tau) = - \int_0^\infty dz \, z \, \frac{\partial}{\partial z} [\, \phi(z,\tau) \, \nu(z,\gamma) \,]$$

$$= \int_0^\infty dz \, \phi(z,\tau) \, \nu(z,\gamma) . \tag{A.44}$$

Here, we integrated by parts and used the fact that $z\phi$ vanishes for $z = 0, \infty$. The quantity $\nu(z,\gamma) > 0$ since the clusters are assumed to grow continuously. Hence,

$$\int_0^\infty dz \ z \ \phi(z,\tau)$$

also continues to grow and the conservation equation (A.30) cannot be fulfilled. Therefore, $\gamma \rightarrow \infty$ is not a possible solution.

b) $\gamma \rightarrow 0$

This case would imply

$$\frac{dx^3}{dt'} \rightarrow \infty \ , \tag{A.45}$$

so that x^3 would grow faster than t'. To check the consistency of this case we use Eq. (A.28) in the form

$$\frac{dz}{d\tau} \simeq - \gamma(\tau) - z \ , \tag{A.46}$$

where we neglected the term $\propto \gamma \ z^{1/3}$ in comparison to z. The solution of Eq. (A.46) with the condition $z(\tau=0) = y$ is

$$z(\tau) = y \ e^{-\tau} - e^{-\tau} \int_0^\tau d\tau' \ e^{\tau'} \ \gamma(\tau') \ . \tag{A.47}$$

Using this result in Eq. (A.38) yields

$$y_0(\tau) = \int_0^\tau d\tau' \ e^{\tau'} \ \gamma(\tau') = 3 \int dx^3 \ \frac{dt'}{dx^3} = 3 \int_0^{t'(\tau)} dt' = 3 \ t'(\tau) \ , \tag{A.48}$$

where Eqs. (A.27) and (A.40) were employed.

To check if this result allows to satisfy the conservation equation (A.37) we study the second term on the RHS of Eq. (A.37),

$$T = \int_{y_0(\tau)}^{\infty} dy \, f_0(y) \, e^{\tau} \, z(y,\tau) \,. \tag{A.49}$$

With the help of Eq. (A.47) we see that

$$e^{\tau} \, z(y,\tau) = y - \int_0^{\tau} d\tau' \, e^{\tau'} \, \gamma(\tau') \le y \,, \tag{A.50}$$

since the integrand is positive. Therefore,

$$T \le \int_{y_0(\tau)}^{\infty} dy \, f_0(y) \, y = \int_{3t'(\tau)}^{\infty} dy \, f_0(y) \, y \,, \tag{A.51}$$

where Eq. (A.48) has been used. As $\tau \to \infty$, $t' \to \infty$, and therefore $T \to 0$, and Eq. (A.37) cannot be satisfied, since also the first term on the RHS goes to zero. Hence, $\gamma \to 0$ is also not a possible solution.

c) $\gamma \to \gamma_0 = $ const.

For this case Eq. (A.29) yields

$$\frac{dx^3}{dt'} = \frac{3}{\gamma_0} \quad \to \quad x(t') = \left[x(t'=0) + \left(\frac{3t'}{\gamma_0} \right) \right]^{1/3}, \tag{A.52}$$

i.e., the asymptotic growth law $x \propto t^{1/3}$. Now, it remains to express the average cluster radius in terms of x. For this purpose we use $\gamma = \gamma_0$ in the continuity equation (A.39),

$$\frac{\partial}{\partial \tau} \phi(z,\tau) + \frac{\partial}{\partial z} [\, \phi(z,\tau) \, v(z,\gamma_0) \,] = 0 \,. \tag{A.53}$$

The ansatz

$$\phi(z,\tau) = - \frac{1}{v(z,\gamma_0)} \chi[\tau + \psi(z)] \tag{A.54}$$

solves this equation if

$$\psi(z) = - \int_0^z dz' \ \frac{1}{v(z',\gamma_0)} \ . \tag{A.55}$$

To obtain further conditions for the function χ we insert the ansatz (A.54) into the conservation equation (A.30),

$$1 = \frac{\Delta_0}{\mathcal{Q}_0} e^{-\tau/3} - \kappa \ e^{\tau} \int_0^{\infty} dz \ \frac{z}{v(z,\gamma_0)} \ \chi[\tau + \psi(z)] \ . \tag{A.56}$$

Since χ must compensate the exponentially divergent prefactor, we make the next ansatz

$$\chi = A \ e^{-\tau} \ e^{-\psi(z)} \ , \tag{A.57}$$

so that

$$\phi(z,\tau) = - \ \frac{A}{v(z,\gamma_0)} \ e^{-\tau} \ e^{-\psi(z)} = - \ e^{-\tau} \ \Phi(z) \ . \tag{A.58}$$

Then Eq. (A.56) yields for large τ,

$$\frac{1}{\kappa} = - A \int_0^{\infty} dz \ \frac{z}{v(z,\gamma_0)} \ e^{-\psi(z)} = \int_0^{\infty} dz \ z \ \Phi(z) \ . \tag{A.59}$$

Before we determine $\Phi(z)$ explicitly, we want to evaluate the cluster growth law.

Inserting Eq. (A.58) into the continuity equation (A.53) yields

$$- \Phi(z) + \frac{\partial}{\partial z} [\ \Phi(z) \ v(z,\gamma_0) \ ] = 0 \ . \tag{A.60}$$

We multiply Eq. (A.60) by z, integrate over all z, and use Eq. (A.59) to obtain

$$- \frac{1}{\kappa} - \int_0^\infty dz \; \Phi(z) \; v(z,\gamma_0) = 0 \; , \tag{A.61}$$

where we partially integrated the second term and inserted $z\Phi = 0$ for $z = 0, \infty$. Using Eq. (A.28) for $v(z,\gamma_0)$, Eq. (A.61) yields

$$\gamma_0 \, (\, \overline{z}^{1/3} - 1) = 0 \; , \tag{A.62}$$

where

$$\overline{z}^{1/3} = \int_0^\infty dz \; z^{1/3} \, \Phi(z) \; . \tag{A.63}$$

Inserting the definition of z, Eq. (A.27), into Eq. (A.62) implies

$$\overline{\rho} = \overline{x} \; . \tag{A.64}$$

Since it follows from Eq. (A.52) that $x(t) \propto t^{1/3}$, Eq. (A.64) with Eq. (A.19) implies

$$\overline{R}(t) \propto t^{1/3} \; . \tag{A.65}$$

This is the celebrated Lifshitz-Slezov growth law for the average cluster radius.

The asymptotic cluster distribution function is also known, Eq. (A.58). The only quantities left to compute are the explicit result for $\psi(z)$ from Eq. (A.55), the normalization constant A, and the numerical value of γ_0. For this purpose Lifshitz and Slezov (1961) investigate the stability of the solution (A.58) for $\gamma = \mathrm{const.} \neq \gamma_0$. They obtain stable solutions which satisfy the coupled Eqs. (A.27), (A.59), and (A.60) only if $\gamma = \gamma_0$ and if $\Phi(z) \equiv 0$ for $z \geq z_0$. z_0 and γ_0 are determined from the conditions,

$$v(z_0,\gamma_0) = 0 = \frac{\partial}{\partial z} \, v(z,\gamma_0) \bigg|_{z \, = \, z_0} . \tag{A.66}$$

The solution of Eq. (A.66) yields

$$\gamma_0 = \frac{27}{4} \quad \text{and} \quad z_0 = \frac{\gamma_0}{2} . \tag{A.67}$$

Using this value for γ_0 in Eq. (A.55) and evaluating the integral gives

$$\psi(z) = \frac{4}{3} \ln(z^{1/3} + 3) + \frac{5}{3} \ln\left[\frac{3}{2} - z^{1/3}\right]$$

$$+ 1 / \left[1 - \frac{2}{3} z^{1/3}\right] - \ln\left(27 \, e \, 2^{-5/3}\right) . \tag{A.68}$$

The normalization constant A is determined from Eq. (A.59). Evaluating the integral yields an expression in terms of an exponential integral. The resulting numerical value for the normalization constant is

$$A \simeq 0.22 \, \frac{Q_0}{R_c^3(0)} . \tag{A.69}$$

From Eq. (A.23) we obtain the number of clusters per unit volume as

$$n_{cluster}(\tau) = \int_0^\infty d\rho^3 \, f(\rho^3, t') = \int_0^\infty dz \, \phi(z, \tau) = A \, e^{-\tau} , \tag{A.70}$$

since the integral over the continuity equation gives

$$\int_0^\infty dz \, \Phi(z) = A . \tag{A.71}$$

Summarizing the results for the asymptotic distribution function, Lifshitz and Slezov (1961) get

$$\phi(z, \tau) = \begin{cases} n_{cluster}(\tau) \, p(z, \gamma_0) & \text{for } z \leq z_0 \\ 0 & \text{for } z \leq z_0 \end{cases} \tag{A.72}$$

where

$$p(z, \gamma_0) = 27 \cdot 2^{-5/3} e \, (z^{1/3} + 3)^{-7/3} (1.5 - z^{1/3})^{-11/3}$$

$$\times \exp\left[-1 \Big/ \left(1 - \frac{2}{3} z^{1/3}\right)\right]. \tag{A.73}$$

Since

$$p(z^{1/3}) \, dz^{1/3} = p(z) \, dz \tag{A.74}$$

we obtain the probability to find a cluster with dimensionless mean radius $z^{1/3} \propto R$, Eqs. (A.27) and (A.19), as

$$p(z^{1/3}) = 3 \, z^{2/3} \, p(z) . \tag{A.75}$$

This asymptotially correct universal size distribution is plotted in Fig. (A.1).

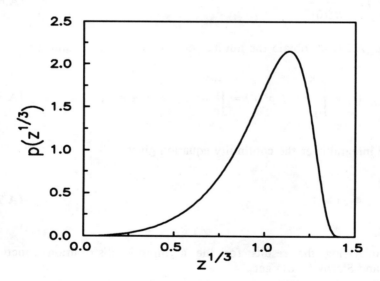

Fig. A.1: Asymptotic cluster size distribution, Eq. (A.75), as function of $z^{1/3} \propto R$.

REFERENCES

Abramowitz, M. and Stegun, I. (eds.) (1970), *Handbook of Mathematical Functions*, Dover Pulications Inc., New York.

Aers, G. C. and Liu, H. C. (1990), Sol. State Comm. **B73**, 19.

Alivisatos, A. P., Harris, A. L., Levinos, N. J., Steigerwald, M. L., and Brus, L. (1988), J. Chem. Phys. **89**, 4001.

Alsmeier, J., Batke, E., and Kotthaus, J. P. (1990), Surf. Sci. **229**, 287.

Anderson, P. W., (1958) Phys. Rev. **109**, 1492.

Bagnal, C.M. and Zarzycki, J. (1990), J. Non-Cryst. Sol. **121**, 221.

Bakshi, P., Broido, D. A., and Kempa, K. (1990), Phys. Rev. **B42**, 7416.

Baldareschi, A. and Lipari, N. O. (1973), Phys. Rev. **B8**, 2697.

Bányai, L. (1964), Physique des Semiconducteurs (*ed*. M. Hulin), p.417, Dunod, Paris

Bányai, L. and Koch, S. W. (1986), Phys. Rev. Lett. **57**, 2722.

Bányai, L., Lindberg, M., and Koch, S. W. (1988a), Opt. Lett. **13**, 212.

Bányai, L., Hu, Y. Z., Lindberg, M., and Koch, S. W. (1988b), Phys. Rev. **B38**, 8142.

Bányai, L., Galbraith, I., and Haug, H. (1988), Phys. Rev. **B38**, 3931.

Bányai, L. (1989), Phys. Rev. **39**, 8022.

Bányai, L., Gilliot, P., Hu, Y. Z., and Koch, S. W. (1992), Phys. Rev. **45B**, 14136.

Baroff, G. A., Kane, E. O., and Schlüter, M. (1979), Phys. Rev. Lett. **43**, 956.

Bastard, G. (1988), *Wave Mechanics Applied to Semiconductor Heterostructures*, Les Editions de Physique, Paris.

Bastard, G., Mendez, E. E., Chang, L. L., and Esaki, L. (1983), Phys. Rev. **B28**, 3241.

Bawendi, M. G., Wilson, W. L., Rothberg, L., Caroll, P. J., Jedju, T. M., Steigerwald, M. L. and Brus, L. E. (1990) Phys. Rev. Lett. **65**, 1623.

Belleguie, L. and Bányai, L. (1991), Phys.Rev. **B44**, 8785

Belleguie, L. and Bányai, L. (1993), Phys.Rev. **B** (to be published).

Bloembergen, N. (1965), *Nonlinear Optics*, Benjamin Inc., New York.

Borelli, N. F., Hall, D. W., Holland, H. J., and Smith, D. W. (1987), J. Appl. Phys. **61**, 5399.

Bottcher, C. F. (1973) Theory of Electric Polarization, 2nd *ed.*, Elsevier, Amsterdam

Brewer, R. (1977), *p.* 341 in *Frontiers in Laser Spectroscopy*, Vol. 1, *ed.* R. Balian, S. Haroche, and S. Liberman, North Holland Publ., Amsterdam.

Brunner, K., Bockelmann, U., Abstreiter, G., Walther, M., Böhm, G., Tränkle, G., and Weimann, G. (1992), Phys. Rev. Lett. **69**, 3216.

Brus, L. E. (1984), J. Chem. Phys. **80**, 4403.

Brus, L. E. (1986), IEEE J. Quantum Electron. QE-22, 1909.

Brus, L.E. (1991), Appl. Phys. **A53**, 465.

Brus, L. E. (1992), 182nd Meeting of the Electrochemical Society, Toronto, Canada, Oct. 11 - 16, 1992.

Bryant, G. W. (1987), Phys. Rev. Lett. **59**, 1140.

Bryant, G. W. (1988), Phys. Rev. **B37**, 8763.

Bryant, G. W. (1989a), Comments Cond. Mat. Phys. **14**, 277.

Bryant, G. W. (1989b), Phys. Rev. **B40**, 1620.

Bryant, G. W. (1990), Phys. Rev. **B41**, 1243.

Bryant, G. W. (1992), Phys. Rev. **B46**, 1893.

Bryant, G. W. (1993), Physica *B* (March), and submitted to Phys. Rev.

Chemla, D. S. and Miller, D. A. B. (1986) Optics Lett. **11**, 522.

Chiba, Y. and Ohnishi, S. (1988), Phys. Rev. **B38**, 12988.

Chiba, Y. and Ohnishi, S. (1989), Superlatt. Microstruct. (UK) **6**, 23.

Cibert, J., Petroff, P. M., Dolan, G. J., Pearton, S. J., Gossard, A. S., and English, J. H. (1986), Appl. Phys. Lett. **49**, 1275.

Cotter, D., Girdlestone, H. P., and Moulding, K. (1991), Appl. Phys. Lett. **58**, 1455.

D'Andrea, A. and Del Sole, R. (1990), Solid State Commun. **74** 1121.

D'Andrea, A., Del Sole, R., Girlanda, R., and Quattropani, A. (eds.) (1992), *Optics of Excitons in Confined Systems*, Institute of Physics Conference Series **123**, Institute of Physics, Bristol.

Dameron, C. T., Reese, R. N., Mehra, R. K., Kortan, A. R., Carroll, P. J., Steigerwald, M. L., Brus, L. E., and Winge, D. R. (1989), Nature **338**, 596.

Darwin, C. G. (1930), Proc. Cambridge Philos. Soc. **27**, 86.

DeLong, K. W., Gabel, A., Seaton, C. T. and Stegeman, G. I., (1989) J. Opt. Soc. Am. **B6**, 1306.

Demel, T., Heitmann, D., Grambow, P., and Ploog, K. (1990a), Phys. Rev. Lett. **64**, 788.

Demel, T., Heitmann, D., Grambow, P., and Ploog, K. (1990b), *p*. 51, in *Localization and Confinement of Electrons in Semiconductors*, Kuchar, F., Heinrich H., and Bauer, G. (eds.), Springer, Berlin.

Desai, R. C., Koch, S. W., and Abraham, F. F. (1983), Physics 118A, 136.

Dingle, R. B. (1952), Proc. Roy. Soc. London A211, 500.

Douglas, T. and Theopold, K. H. (1991), Inorg. Chem. 30, 594.

Edmonds, A. R. (1957), *Angular Momentum in Quantum Mechanics*, Princeton University Press, Princeton, NJ.

Efros, Al. L. and Efros, A. L. (1982), Sov. Phys. Semicond. 16, 772.

Efros, Al. L. and Rodina, A. V. (1989) Solid State Comm. 72, 645

Ekimov, A. I. and Efros, A. L. (1988), phys. stat. sol. b150, 498.

Ekimov, A. I., Efros, Al. L., and Onuschenko, A. A. (1985), Solid State Comm. 56, 921.

Ekimov, A. I. and Onuschenko, A. A. (1981), JETP Lett. 34, 345.

Ekimov, A. I. and Onuschenko, A. A. (1982), Sov. Phys. Semicond. 16, 775.

Ekimov, A. I., Skvortsov, A. P., Shubina, T. V., Shumilov, S. K. and Efros, Al. (1989) Sov. Phys. Tech. Phys. 34, 371

Esch, V., Fluegel, B., Khitrova, G., Gibbs, H. M., Jiajin, X., Kang, K., Koch, S. W., Liu, L. C., Risbud, S. H., and Peyghambarian, N. (1990), Phys. Rev. B42, 7450.

Fock, V. (1928), Z. Phys. 47, 446.

Galeuchet, Y. D., Rothuizen, H., and Roentgen, P. (1991), Appl. Phys. Lett. 58, 2423.

Garnett, M. (1904), Philos. Trans. R. Soc. London 203, 385.

Gilliot, P., Merle, J. C., Levy, R., Robino, M., and Hönerlage, B. (1989), phys. stat. sol. b153, 403.

Gilliot, P., Hönerlage, B., Levy, R., and Grun, J. B. (1990), phys. stat. sol. b159, 259.

Grätzel, M. (1989), Nature 338, 5401.

Hache, F., Ricard, D., and Flytzanis, C. (1989), Appl. Phys. Lett. 55, 1504.

Hanamura, E. (1987), Solid State Comm. 62, 465.

Hanamura, E. (1988), Phys. Rev. B37, 1273

Hanamura, E., Kuwata-Gonokami, M., and Ezaki, H. (1990), Solid State Comm. 73, 551.

Hansen, W., Smith, T. P., Lee, K. Y., Brum, J. A., Knoedler, C. M., Hong, J. M., and Kern, D. P. (1988), Phys. Rev. B38, 2172.

Haug, H. and Bányai L. (eds.) (1989), *Optical Switching in Low-Dimensional Systems*, Plenum Press, New York.

Haug, H. and Koch, S. W. (1990, 1993), *Quantum Theory of the Optical and Electronic Properties of Semiconductors*, World Scientific, Singapore; second edition (1993).

Haug, H. and Schmitt-Rink, S. (1984), Prog. Quant. Electr. 9, 3.

Henneberger, F., Woggon, U., Puls, J., and Spiegelberg, Ch. (1988), Appl. Phys. B46, 19.

Hilinski, E. F., Lucas, P. A., and Wang, Y. (1988), J. Chem. Phys. 89, 3435.

Hu, Y. Z. (1991), Ph.D. dissertation (University of Arizona).

Hu, Y. Z., Koch, S. W., Lindberg, M., Peyghambarian, N., Pollock, R., and Abraham, F. F. (1990a), Phys. Rev. Lett. 64, 1805.

Hu, Y. Z., Lindberg, M., and Koch, S. W. (1990b), Phys. Rev. B42, 1713.

Hu, Y. Z., Koch, S. W., and Tran Thoai, D. B. (1990c), Mod. Phys. Lett. B4, 1009.

Huang, K. (1950), A. Rhys. Proc. R. Soc.(London) A204, 406.

Huang, D. and Antoniewicz, P. R. (1991), Phys. Rev. B43, 2169.

Huang, D. and Zhou, S. (1989), Phys. Rev. B40, 8235.

Ishihara, H. and Cho, K. (1990) Phys. Rev. B42, 1724

Kane, E. O. (1957), J. Phys. Chem. Solids 1, 249.

Kang, K. I., McGinnis, B. P., Sandalphon, Hu, Y. Z., Koch, S. W., Peyghambarian, N., Mysyrowicz, A., Liu, L. C., and Risbud, S. H. (1992), Phys. Rev. B45, 3465 (1992).

Kash, K. (1990), J. of Luminescence 46, 69.

Kash, K., Bhat, R., Mahoney, D. D., Lin, P. S. D., Scherer, A., Warlock, J. M., Van der Gaag, B. P., Koza, M., and Grabbe, P. (1989), Appl. Phys. Lett. 55, 681.

Kash, K., Van der Gaag, B. P., Worlock, J. M., Gozdz, A. S., Mahoney, D. D., Harbison, J. P., and Florez, L. T. (1990), p. 39, in Localization and Confinement of Electrons in Semiconductors, Kuchar, F., Heinrich, H., and Bauer, G. (eds.), Springer, Berlin.

Kash, K., Scherer, A., Worlock, J. M., Craighead, H.G., and Tamargo, M. C. (1986), Appl. Phys. Lett. 49, 1043.

Kastner, M. A. (1993), Physics Today 46, No. 1, 24.

Kayanuma, Y. (1986), Solid State Commun. 59, 405.

Kayanuma, Y. (1993), J. Phys. Soc. Japan 62, No. 1 (January 1993).

Kern, K., Heitmann, D., Grambow, P., Zhang, Y. H., and Ploog, K. (1991), Phys. Rev. Lett. 66, 1618.

Kippelen, B., Levy, R., Faller, P., Gilliot, P., Belleguie, L. (1991), Appl. Phys. Lett. 59, 3378.

Klein, M. C., Hache, F., Richard, D., and Flytzanis, C. (1990), Phys. Rev. B42, 11123.

Kleinman, D. A. (1988), Phys. Rev. B28, 871.

Kobayashi, M., Ohno, Y., Endo, S., Cho, K., and Narita, S. (1983), in Proceedings of the 16th International Conference on the Physics of Semiconductors, M. Averous (ed.), North-Holland, Amsterdam.

Koch, S. W. (1984), Dynamics of First-Order Phase Transitions in Equilibrium and Nonequilibrium Systems, Lecture Notes in Physics 207, Springer, Berlin.

Koch, S. W. (1990), Phys. Bl. **46**, 167.

Koch, S. W., Desai, R. C., and Abraham, F. F. (1983), Phys. Rev. **A27**, 2151.

Koch, S. W., and Liebmann, R. (1983), J. Stat. Phys. **33**, 31.

Koch, S. W., Hu, Y. Z., and Peyghambarian, N. (1992), J. Crystal Growth **117**, 592.

Koch, S. W., Hu, Y. Z., and Binder, R. (1993), Physica *B* (March 1993).

Kotthaus, J., Lorke, A., Alsmeier, J., and Merkt, U. (1990), *p.* 29, in *Localization and Confinement of Electrons in Semiconductors*, Kuchar, F., Heinrich, H., and Bauer, G. (eds.), Springer, Berlin.

Kouwenhoven, L. P., van Wees, B. J., Hekking, F. W. J., Harmans, K. J. P. M., and Timmering, C. E. (1990), *p.* 77, in *Localization and Confinement of Electrons in Semiconductors*, Kuchar, F., Heinrich, H., and Bauer, G. (eds.), Springer, Berlin.

Kranz, H. H. and Haug, H. (1986), J. Lumin. **34**, 337.

Kuchar, F., Heinrich, H., and Bauer, G. (eds.) (1990), *Localization and Confinement of Electrons in Semiconductors*, Springer, Berlin.

Kumar, A., Laux, S. E., and Stern, F. (1990), Phys. Rev. **B42**, 5166.

Kunath, W., Zemlin, F., and Weiss, K. (1985). Ultramicroscopy **16**, 123. (1985).

Lent, C. S. (1991), Phys. Rev. **B43**, 4179.

Leung, K. M. (1986) Phys. Rev. **A33**, 2461

Lifshitz, I. M. and Slezov, V. V. (1959), Soviet Physics JETP **35**, 331.

Lifshitz, I. M. and Slezov, V. V. (1961), J. Phys. Chem. Sol. **19**, 35.

Lindblad, G. (1976), Commun. Math. Phys. **48**, 119.

Lippens, P. E. and Lannoo, M. (1989), Phys. Rev. **B39**, 10935.

Liu, L. C. and Risbud, S. H. (1990), J. Appl. Phys. **68**, 28 (1990).

Lorke, A., Kotthaus, J., and Ploog, K. (1990), Phys. Rev. Lett. **64**, 2559.

Luttinger, J. M. (1956), Phys. Rev. **102**, 1030.

Ma, H., Gomes, S. L., and de Araujo C. B. (1992), Journ. Opt. Soc. Am. **B9**, 2230 (1992).

MacKenzie, J.D. (1992), Proc. SPIE **1758**, Conf. on Sol-Gel Optics II, July 20 - 22, San Diego, CA.

Maeda, Y., Tsukamoto, N., Yazawa, Y., Kanemitsu, Y., and Masumoto, Y. (1991), Appl. Phys. Lett. **59**, 3168.

Maksym, P. A. and Chakraborty, T. (1990), Phys. Rev. Lett. **65**, 108.

Malhotra, J., (1988) Thesis, University of Orlando, Florida.

Masumoto, Y., Yamazaki, M., and Sugawara, H. (1989), Appl. Phys. Lett. **53**, 1527.

Masumoto, Y., Ideshita, S., and Wamura, T., (1990), phys. stat. sol. **b159**, 133.

Merkt, U. (1990), Adv. in Solid State Physics **30**, 77.

Milanovic, V. and Ikonic, Z. (1989) Phys. Rev. **B39**, 7982

Miller, D. A. B., Chemla, D. S., Damen, T. C., Gossard, A. C., Wiegmann, W., Wood, T. H., and Burrus, C. A. (1985), Phys. Rev. **B32**, 1043.

Morgan, R. A., Park, S. H., Koch, S. W., and Peyghambarian, N. (1990), Semicond. Sci. Technol. **5**, 544.

Mott, N. F. (1966), Phil Mag. **13**, 989

Mott, N. F. (1968) J. Non-Cryst. Solids **1**, 1.

Nair, S. V., Sinha, S., and Rustagi, K. C. (1987), Phys. Rev. **B35**, 4098.

Nomura, S. and Kobayashi, T. (1990a), Solid State Commun. **73**, 425.

Nomura, S. and Kobayashi, T. (1990b) Solid State Commun. **74**, 1153.

Nozue, Y., Tang, Z. K., and Goto, T. (1990), Sol. State Comm. **73**, 531.

Olshavsky, M. A., Goldstein, A. N., and Alivisatos, A. P. (1990), J. Am. Chem. Soc. 112, 9438.

Park, S. H., Morgan, R. A., Hu, Y. Z., Lindberg, M., Koch, S. W., and Peyghambarian, N. (1990), J. Opt. Soc. Am. B7, 2097.

Peeters, F. M. (1990), Phys. Rev. B42, 1486.

Pettifor, D. G. and Weaire, D. L. (eds.), (1985), *The Recursion Method and its Applications*, Springer Verlag, Berlin.

Peyghambarian, N. and Koch, S. W. (1990), *p.* 7 in: *Nonlinear Photonics*, eds. H.M. Gibbs *et al.*, Springer Verlag, Berlin.

Peyghambarian, N., Fluegel, B., Hulin, D., Migus, A., Joffre, M., Antonetti, A., Koch, S. W., and Lindberg, M. (1989), IEEE J. Quantum Electronics 25, 2516.

Pollok, E. L. (1988), Comput. Phys. Commun. 52, 49.

Pollock, E. L. and Ceperley, D. M. (1984), Phys. Rev. B30, 2555.

Pollock, E. L. and Koch, S. W. (1991), J. Chem. Phys. 94, 6766.

Pollock, E. L. and Runge, K. J., (1992) "Lectures on Path-Integral Computations", Lawrence Livermore National Lab. preprint.

Que, W. and Kirczenow, G. (1988), Phys. Rev. B38, 3614.

Ramsden, J. J. and Graetzel, M. (1984), J. Chem. Soc. Faraday Trans. *I* 80, 919.

Reed, M. A. (1993), Scientific American (January 1993), 118.

Reed, M. A., Bate, R. T., Bradshaw, K., Duncan, W. M., Frensley, W. R., Lee, J. W., and Shih, H. D. (1986) J. Vac. Sci. Technol. B4, 1.

Reed, M. A., Randall, J. N., Aggarwal, R. J., Matyi, R. J., Moore, T. M., and Wetsel, A. E. (1988), Phys. Rev. Lett. 60, 535.

Reed, M. A., Randall, J. N., and Luscombe, J. H. (1990), in *Localization and Confinement of Electrons in Semiconductors*, Kuchar, F., Heinrich H., and Bauer, G. (eds.), Springer, Berlin.

Rosetti, R., Ellison, J. L., Gibson, J. M., and Brus, L. E. (1984), J. Chem. Phys. **80**, 4464.

Rotenberg, M., Bivins, R., Metropolis, N., and Wooten, J. K. (1959), *The 3-j and 6-j Symbols*, Technology Press, Cambridge, Mass.

Roussignol, P., Ricard, D., Lukasik, J., and Flytzanis, C. (1987), J. Opt. Soc. Am. **B4**, 5.

Roussignol, P., Ricard, D., Flytzanis, C., and Neuroth, N. (1989), Phys. Rev. Lett. **62**, 312.

Schanne-Klein, M. C., Hache, F., Ricard, D., and Flytzanis, C. (1992), J. Opt. Soc. Am. **B9**, 2234.

Schiff, L. I. (1968), *Quantum Mechanics*, McGraw-Hill, New York.

Schmitt-Rink, S., Miller, D. A. B., and Chemla, D. S. (1987), Phys. Rev. **B35**, 8113.

Sercel, P. C. and Vahala, K. J. (1990), Phys. Rev. **B42**, 3690.

Sheu, S. Y., Kummar, S., and Cukier, R. J. (1990), Phys. Rev. **B42**, 431.

Sikorski, C. and Merkt, U. (1989), Phys. Rev. Lett. **62**, 2164.

Sikorski, C. and Merkt, U. (1990) Surf. Sci. **229**, 282.

Sivan, U., Imry, Y. and Hartzstein, C. (1989) Phys. Rev. **B39**, 1242

Smith III, T. P. (1990) in *Localization and Confinement of Electrons in Semiconductors*, Kuchar, F., Heinrich, H., and Bauer, G. (eds.), Springer, Berlin, p.10.

Sommerfeld, A. (1949) *Elektrodynamik*, Akademische Verlagsgesellschaft Geest & Portig K.-G., Leipzig.

Spano, F. C., Mukamel, S. (1989), Phys. Rev. **A40**, 5783

Stenholm, S. (1977), *p.* 399 in *Frontiers in Laser Spectroscopy*, Vol. 1, *ed.* R. Balian, S. Haroche, and S. Liberman, North Holland Publ., Amsterdam.

Stolz, G. (1989), preprint (in German), Mathematics Institute, University Frankfurt, Germany.

Sweeny, M. and Xu, J. (1989), Solid State Commun. **72**, 301.

Takagahara, T. (1988), Phys. Rev. **B39**, 10206.

Takagahara, T. and Takeda, K. (1992), Phys. Rev. **B46**, 15578.

Tang, Z. K., Nozue, Y. and Goto, T. (1991), J. Phys. Soc. Jpn. **60**, 2090.

Temkin, H., Dolan, G.J., Panish, M. B., and Chu, S. N. G. (1987), Appl. Phys. Lett. **50**, 413.

Tomita, M. and Matsuoka, M. (1990) J. Opt. Soc. Am. **B7**, 1198.

Tommasi, R., Lepore, M., Ferrara, M., and Catalano, I. M. (1992), Phys. Rev. **B46**, 12261.

Toyozawa, Y. (1983), in *Proceedings of the 16th International Conference on the Physics of Semiconductors*, M. Averous (*ed*), North-Holland, Amsterdam.

Tran Thoai, D. B., Hu, Y. Z., and Koch, S. W. (1990a), Phys. Rev. **B42**, 4137.

Tran Thoai, D. B., Zimmermann, R., Grundmann, M., and Bimberg, D. (1990b), Phys. Rev. **B42**, 5906.

Tsuboi, T. (1980), J. Chem. Phys. **72**, 5343.

Tsuchiya, M., Gaines, J. M., Yan, R. H., Simes, R. J., Holtz, P. O., Coldren, L. A., and Petroff, P. M. (1989), Phys. Rev. Lett. **62**, 466.

Uchida *et al.* (1992)

Uhrig, A., Bányai, L., Hu, Y. Z., Koch, S. W., Klingshirn, C., and Neuroth, N. (1990), Z. Physik **B81**, 385.

Vandichev, U. V., Dneprovski, V. S., Ekimov, A. I., Okorokov, D. K., Popova, L. B., and Efros, Al. L. (1987), JETP Lett. **46**, 495.

Vahala, K. J. (1990), in: Proc. SPIE OE'LASE 90 Conf. Los Angeles, Jan. 14-19 1990, Paper 1216-13.

Vahala, K. J. and Sercel, P. C. (1990), Phys. Rev. Lett. **65**, 239.

Wang, Y. and Herron, N. (1990), Phys. Rev. **B42**, 7253.

Wang, Y., Herron, N., Mahler, W., and Sune, A. (1989), J. Opt. Soc. Am. **B6**, 808.

Warnock, J. and Awschalom, D. D. (1985), Phys. Rev. **32**, 5529.

Warnock, J. and Awschalom, D. D. (1986), Appl. Phys. Lett. **48**, 425.

Weller, H., Schmidt, H. M., Koch, U., Fojtik, A., Baral, S., Henglein, A., Kunath, W., Weiss, K., and Dieman, E. (1986), Chem. Phys. Lett. **124**, 557.

Weyl, H. (1910), Math. Ann. **68**, 220.

Williams, V. S., Olbright, G. O., Fluegel, B., Koch, S. W., and Peygham-barian, N. (1988), J. Mod. Opt. **35**, 1979.

Woggon, U., Müller, M., Lembke, U., Rückmann, I., and Cesnulevicius, J. (1991), Superlatt. Microstruct. **9**, 245.

Woggon, U., Gaponenko, S., Langbein, W., Uhrig, A. and Klingshirn, C. (1993) (submitted to Phys. Rev. **B**).

Wu, W. Y., Schulman, J. N., Hsu, T. Y., and Efron, U. (1987), Appl. Phys. Lett. **51**, 710.

Xia, J. B. (1989), Phys. Rev. **B40**, 8500.

Zarem, H. A., Sercel, P. C., Hoenk, M. E., Lebens, J. A., and Vahala, K. J. (1989), Appl. Phys. Lett. **54**, 2692.

Zhu, Y., Huang, D., and Feng, S. (1989), Phys. Rev. **B40**, 3169.

Zimin, L. G., Gaponenko, S. V., Lebed, V. Yu., Malinovskii, I. E., Germanenko, I. N., Podorova, E. E., and Tsekhomskii, V. (1990a), phys. stat. sol. **b159**, 111.

Zimin, L. G., Gaponenko, S. V., Lebed, V. Yu., Malinovskii, I. E., and Germanenko, I. N. (1990b), J. of Luminescence **46**, 101.

Zimmermann, R. (1988), *Many-Particle Theory of Highly Excited Semiconductors*, BSB Teubner, Leipzig.

INDEX